MYANMAR

the state, community and the environment

MYANMAR

the state, community and the environment

Monique Skidmore and Trevor Wilson (eds)

ANU
THE AUSTRALIAN NATIONAL UNIVERSITY

E PRESS

Asia Pacific Press
The Australian National University

ANU

E PRESS

Copublished by ANU E Press and Asia Pacific Press
The Australian National University
Canberra ACT 0200 Australia
Email: anuepress@anu.edu.au
This title is available online at http://epress.anu.edu.au/myanmar_citation.html

Asia Pacific Press
Crawford School of Economics and Government
The Australian National University
Canberra ACT 0200
Ph: 61-2-6125 0178 Fax: 61-2-6125 0767
Email: books@asiapacificpress.com
Website: http://www.asiapacificpress.com

National Library of Australia Cataloguing in Publication entry

---

Myanmar : the state, community and the environment.

> Bibliography.
> Includes index.
> ISBN 9780731538119 (pbk.)
> ISBN 9781921313370 (online)
>
> 1. Burma - Economic conditions. 2. Burma - Environmental
> conditions. 3. Burma - Politics and government. 4. Burma -
> Social conditions. I. Wilson, Trevor. II. Skidmore,
> Monique. III. Title.
>
> 320.9591

Cover design: ANU E Press
The cover image is based on the oil painting 'Welcome Summer' by Shwe Maung Thar, a Rakhine artist living in Mrauk-U, western Myanmar.

Contents

Tables   vi
Figures   vii
Maps   vii
Abbreviations   viii
Contributors   xi
Editors' note   xvi

Introduction   xvii

**Assessing political/military developments after the departure
of Khin Nyunt**

1. The political situation in Myanmar   1
   Vicky Bowman
2. A Burmese perspective on prospects for progress   18
   Khin Zaw Win
3. Of *kyay-zu* and *kyet-su*: the military in 2006   36
   Mary Callahan
4. Conflict and displacement in Burma/Myanmar   54
   Ashley South
5. Foreign policy as a political tool: Myanmar 2003–2006   82
   Trevor Wilson

Assessing the economic situation after the 2001–2002 banking crisis

6. Myanmar's economy in 2006   108
   Sean Turnell
7. Transforming Myanmar's rice marketing   135
   Ikuko Okamoto
8. Industrial zones in Burma and Burmese   159
   labour in Thailand
   Guy Lubeigt

Implications of current development strategies for
Myanmar's environment

9. Environmental governance in the SPDC's Myanmar 189
   Tun Myint
10. Environmental governance of mining in Burma 218
    Matthew Smith
11. Spaces of extraction: governance along the riverine 246
    networks of Nyaunglebin District
    Ken MacLean
12. Identifying conservation issues in Kachin State 271
    Tint Lwin Thaung
Index 290

Tables

1.1	The SPDC's seven-step road-map of 30 August 2003	2
1.2	Chapters of the Draft Constitution	4
4.1	Typology of forced migration	56
6.1	Claimed annual GDP growth rates, 1999–2005	110
6.2	Economic growth estimates, 2001–2006	111
6.3	State share of Myanmar's financial resources, selected indicators, 1999–2006	114
6.4	Customs duty revenues, 2002–2006	114
6.5	Indicative (unofficial) exchange rates, 1997–2006	117
6.6	Myanmar's external sector, selected indicators, 1999–2005	120
6.7	Composition of exports, 2002–2005	120
6.8	Foreign direct investment flows, 2003–2005	122
6.9	Foreign exchange reserves, selected countries, 2000–2005	125
6.10	Selected financial indicators, 1999–2006	127
7.1	Estimated volume of domestically marketed rice, 1970–2001	138
7.2	Changes in milled rice by MAPT-owned and MAPT-contracted mills, 1988–2001	141

7.3	Changes in volume of rationed rice	142
7.4	Direction of Myanmar's rice exports, 1990–2001	144
7.5	Number of private mills registered with MAPT	147
7.6	Number of private mills contracted for procurement of paddy, 1999–2001	148
8.1	Development of the private sector in 2006	161
8.2	Industrial zones in Burma, 2006	166
8.3	Labour force and unemployment rate, 1996–2001	168
8.4	Illegal migrant labour from Burma in Thailand	169
8.5	Types of jobs and salaries paid to Burmese workers in Thailand	177
9.1	Current major environmental legislation	198
9.2	International environmental conventions ratified or signed by Burma	202
11.1	Selected list of rents extracted in the mining concessions, Shwegyin Township, 2004–2005	260
12.1	Categories of stakeholders interviewed	272
12.2	Large-scale issues arising from official projects withinstitutional support and driven by larger commercial interests	273
12.3	Small-scale issues, often driven by financial hardship	274
12.4	Threats ranked by stakeholders, regardless of scale	276
12.5	Official logging companies in northern Myanmar	286

Figures

7.1	Changes in procurement and farm-gate prices, 1999–2002	137
11.1	Shwegyin Township Mining Area	247

Maps

8.1	Industrial zones in Burma	160
8.2	Industrial zones along the Thailand-Burma border	166
8.3	Asian highways across Burma	173

Abbreviations

ABSDF	All Burma Students' Democratic Front
ACMECS	Ayeyawaddy (Irrawaddy)-Chao-Phraya-Mekong Economic Cooperation Strategy
ADB	Asian Development Bank
AMD	acid mine drainage
ANU	The Australian National University
ARD	acid rock drainage
ASEAN	Association of Southeast Asian Nations
ASEM	Asia-Europe Meeting
ASM	artisanal and small-scale mining
BBC	British Broadcasting Corporation
BP	British Petroleum
BSPP	Burma Socialist Programme Party
BSS	Burma Selection System
CBD	Convention on Biological Diversity
CBM	Central Bank of Myanmar
CBO	community-based organisation
CITES	Convention on International Trade in Endangered Species of Wildlife and Fauna
COHRE	Centre on Housing Rights and Evictions
CWS	Chatthin Wildlife Sanctuary
DKBA	Democratic Kayin Buddhist Army
DSI	Defence Services Intelligence
ECS	Economic Cooperation Strategy
EIA	environmental impact assessment
EITI	Extractive Industries Transparency Intiative
EIU	Economist Intelligence Unit
EMS	Environmental Management Systems
ERI	Earth Rights International
FAO	Food and Agriculture Organization
FATF	Financial Action Task Force

FCCC	Framework Convention on Climate Change
FDI	foreign direct investment
GAIL	Gas Authority of India Limited
GDP	gross domestic product
GMS	Greater Mekong Subregion
GRI	Global Reporting Initiative
ICG	International Crisis Group
ICRC	International Committee of the Red Cross
IDCE	International Development, Community, and the Environment
ILO	International Labour Organization
IMF	International Monetary Fund
INGO	international non-governmental organisation
ISEAS	Institute of Southeast Asian Studies
ITTA	International Tropical Timber Agreement
KESAN	Karen Environmental and Social Action Network
KHRG	Karen Human Rights Group
KIO	Kachin Independence Organisation
KNU	Karen National Union
KNLA	Karen National Liberation Army
MADB	Myanmar Agricultural Development Bank
MAPT	Myanmar Agricultural Produce Trading
MCSO	Myanmar Central Statistical Office
ME1	Number 1 Myanmar Enterprise
ME2	Number 2 Myanmar Enterprise
ME3	Number 3 Myanmar Enterprise
MGE	Myanmar Gems Enterprise
MICCL	Myanmar Ivanhoe Copper Company Limited
MNDAA	Myanmar National Democratic Alliance Army
MOGE	Myanmar Oil and Gas Enterprise
MPE	Myanmar Pearl Enterprise
MRC	Mekong River Commission
MSF	Médecins sans Frontiéres
MWAF	Myanmar Women's Affairs Federation
NCEA	National Commission for Environmental Affairs (Burma)

NDAK	New Democratic Army—Kachin
NEP	National Environmental Policy
NGO	non-governmental organisation
NLD	National League for Democracy
NMSP	New Mon State Party
NTFP	Non-Timber Forest Products
OECD	Organisation for Economic Co-operation and Development
ONGC	Oil and Natural Gas Corporation
OSI	Open Society Institute
PKDS	Pan Kachin Development Society
PRA	participatory rural appraisal
REFS	Review of the Financial, Economic and Social Conditions
SPDC	State Peace and Development Council
SLORC	State Law and Order Restoration Council
TBBC	Thailand Burma Border Constortium
UMEHL	Union of Myanmar Economic Holdings Limited
UNCED	United Nations Conference on Environment and Development
UNEP	United Nations Environment Programme
UNDP	United Nations Development Programme
UNHCR	United Nations High Commissioner for Refugees
UNICEF	United Nations Childrens Fund
UNIRIN	United Nations Integrated Regional Information Networks
UNODC	United Nations Office on Drugs and Crime
USDA	Union Solidarity and Development Association
WCS	Wildlife Conservation Society
WFP	World Food Program
WWF	World Wildlife Fund

Contributors

Vicky Bowman received a BA (Hons) in Natural Sciences (Pathology) from Pembroke College, Cambridge, before winning a scholarship to the University of Chicago, where she took courses in Latin American Studies. She entered the East Africa Department of the Foreign and Commonwealth Office in 1988 and studied Burmese at London's School of Oriental and African Studies, before being posted as Second Secretary to the British Embassy in Rangoon from 1990–1993. From 1996 to 1999, she was First Secretary/spokeswoman at the British European Union Mission in Brussels, before moving to the European Commission in 1999, where she was a member of the Cabinet of External Affairs Commissioner, Chris Patten, until 2002. From 2002 to 2006, she was British Ambassador to Myanmar. She has published translations of various Burmese short stories, a translation of Mya Than Tint's *On the Road to Mandalay*, and contributed to editions of Lonely Planet's *Burmese Phrasebook*.

Mary P. Callahan is an associate professor of International Studies at the University of Washington. She is author of *Making Enemies: war and state-building in Burma* (2003), which received the Harry J. Benda Prize in Southeast Asian Studies in 2006. Author of numerous articles on modern Burmese politics, Callahan's current research looks at relations between the international community and Burma, the privatisation of security in Southeast Asia and comparative civil–military relations. She received her PhD in Political Science from Cornell University in 1996 and has taught at the US Naval Postgraduate School in Monterey, California.

Guy Lubeigt is a geographer who graduated from the Sorbonne University and National Institute of Oriental Languages and Civilisations. He obtained his PhD in Tropical Geography in 1975 for *Le palmier à sucre en Birmanie Centrale* (Sugar palm tree in Central Burma: culture and exploitation). In 2001, he received his PhD in Asian Studies at the Sorbonne University for *Birmanie: un pays modelé par le Bouddhisme. Essai de géographie religieuse et politique* (Burma: a country modelled by Buddhism: religious and political geography). Now Senior Field Researcher in the National Scientific Research Centre, he is also a member of the Doctoral School of Geography of

Sorbonne University and Visiting Professor at the Southeast Asian Studies Centre of the Chulalongkorn University in Bangkok. He has lived in Burma since 1968. He has specialised in the geography of Buddhism in Burma and has authored 13 books and 55 articles on Burma and Thailand. His publications on religion include *Pagan: histoire et légendes* (1998) and *La Birmanie: l'âge d'or de pagan* (2005).

Ken MacLean holds a PhD in Anthropology and a MSc in Environmental Justice, both from the University of Michigan, Ann Arbor. He is an Assistant Professor in the Department of International Development, Community, and Environment (IDCE) at Clark University. During 2001 and 2002, he served as the Associate Director of EarthRights International's Burma Project. Since then, he has worked as a consultant for EarthRights International on a wide range of research projects. MacLean is the author of numerous research monographs, academic articles, policy briefs submitted to the International Labour Organization and other publications related to contemporary Burma.

Tun Myint is a Research Associate at the Workshop in Political Theory and Policy Analysis at Indiana University, Bloomington. He left Burma after the military coup on 18 September 1988, which cracked down on the people's movement for democracy in which he was involved as a student activist. He came to Indiana University in 1993 after he was awarded a scholarship by the US Information Agency-funded Burmese Refugee Scholarship Program, administered by the Office of International Programs at Indiana University. At Indiana University, he graduated with a BA in Political Science with Honours and East Asian Studies as a double-major degree in 1997, completed a Masters of Public Affairs in 1999 from the School of Public and Environmental Affairs and a PhD in Law and Social Sciences jointly conducted at the School of Public and Environmental Affairs and Indiana University School of Law. He is also a Research Fellow at the Institutional Dimension of Global Environmental Change program, a core science project of the International Human Dimension Program, which strives to achieve scientific understanding of the human dimensions of global environmental change.

Ikuko Okamoto joined the Institute of Developing Economies, Japan, in 1992, after completing a masters degree at the Food Research Institute, Stanford University. She completed a doctoral degree at Kyoto University in 2006, where her PhD topic was 'A study on economic disparity in rural Myanmar: focusing on pulse production after market liberalisation'. Her current major research interests are agricultural and rural development and transitional economies, particularly Myanmar's economy. She is the author of a number of articles in Japanese and English on Myanmar's rural economy.

Matthew Smith is a Project Coordinator for the Burma Project with EarthRights International, focusing on the social and environmental impacts of oil and gas development and the mining of natural resources in Burma. He received a BA in Political Science from Le Moyne College in Syracuse, New York, and a MA in Human Rights Studies and Religion from Columbia University in New York City. During graduate studies, he held an internship at the Early Warning Analysis and Contingency Planning Unit at the United Nations Office for the Coordination of Humanitarian Affairs in New York. Previously, he worked with the Jesuit Volunteer Corps as an emergency services social worker with low-income communities in Alabama, and he has experience in grassroots organising in East Harlem, New York. Before joining EarthRights International, he worked under Kerry Kennedy for Speak Truth to Power, a project of the Robert F. Kennedy Center for Human Rights.

Monique Skidmore is Associate Dean, Postgraduate, College of Arts and Social Sciences and a Fellow in the Research School of Humanities, The Australian National University (ANU). After graduating in anthropology from the ANU, she completed her Masters and PhD degrees at McGill University, Canada. She is the author of many articles and book chapters on Myanmar. Examples of some recent publications are: *Burma at the Turn of the Twenty-First Century* (2005), *Karaoke Fascism: Burma and the politics of fear* (2004), *Women and the Contested State: religion, violence and agency in South and Southeast Asia* (2007, University of Notre Dame Press) and *Medicine in Myanmar: past and present* (2007, NIAS Press). Her research interests include medical anthropology and peace and conflict studies in Southeast Asia.

Ashley South is an independent consultant and analyst, specialising in ethnic politics and humanitarian issues in Burma/Myanmar. He is the author of a political history of lower Burma, *Mon Nationalism and Civil War in Burma: the golden sheldrake* (RoutledgeCurzon; reprint edition 2005), and of an influential essay on the strategic roles of civil society in promoting democratisation, 'Political transition in Myanmar: a new model for democratization' (*Contemporary Southeast Asia*, 26[2]). Research for this paper was conducted during consultancies for the Thailand Burma Border Consortium (2002), International Crisis Group (2003), Human Rights Watch (2004–05), United Nations Development Programme (2005), and with a grant from the John D. and Catherine T. MacArthur Foundation (2003–04). He has a Masters degree in Asian Politics from the School of Oriental and African Studies (London University).

Tint Lwin Thaung, a native of Burma/Myanmar, was trained as a forester, natural resource manager and restoration ecologist in Burma, Thailand and Australia. He has worked on natural resource conservation and community development in Burma/Myanmar and Australia for 20 years. From 1993 to 1997, he worked for the Wildlife Conservation Service in Myanmar. He is dedicated to promoting conservation and development assistance in Burma/Myanmar and to providing training opportunities for younger generations from Myanmar. He has degrees in Forestry from Rangoon University (1985), a Masters degree in Natural Resources from the Asian Institute of Technology, Bangkok (1992), and a PhD from the University of Queensland (2002). He currently lives in Australia. He has published numerous articles on conservation based on fieldwork undertaken in Myanmar.

Sean Turnell is an economist and former central banker with a long-time interest in Burma's financial system. He is based at Macquarie University in Sydney, where he is Senior Lecturer in Economics. Together with colleagues at Macquarie University, in 2001, he founded *Burma Economic Watch*, an online journal of commentary and analysis of Burma's economy. His primary research focus is on Myanmar's financial institutions. He is currently completing a book on Burma's monetary and financial system for the Nordic Institute of Asian Studies. He has lectured on Burma all over the world and, in 2006, he was invited to testify on the country's economy before the US Senate.

Trevor Wilson is a Visiting Fellow in the Department of Political and Social Change, The Australian National University, Canberra. He retired in August 2003 after working for more than 36 years in the Australian government, 30 years of which was spent with the Department of Foreign Affairs and Trade. He served as Australia's Ambassador to Myanmar for three years from mid 2000 to mid 2003. He was assigned to Australia's embassy in Tokyo three times: first in the late 1960s, then in the early 1980s, and finally as Deputy Head of Mission in the second half of the 1990s. He also had tours of duty in Washington and Laos. In Canberra, he also worked in the Defence Department, the Prime Minister's Department and the office of the Minister for Foreign Affairs, Gareth Evans. Since 2004, he has been co-convener of the Myanmar/Burma Update conference series for the ANU and edited *Myanmar's Long Road to National Reconciliation* (2006).

Khin Zaw Win is a citizen of Myanmar. He was educated at schools in Yangon, New Delhi, Madras and Colombo, before training as a dental surgeon in Yangon, graduating in 1971. From 1973 to 1979, he served in the Department of Health, Myanmar; from 1980 to 1983, in the Ministry of Health, Sabah, Malaysia; and, from 1991 to 1992, he was a consultant for UNICEF in Yangon. He attended the Master in Public Policy program at the Centre for Advanced Studies (now the Lee Kwan Yew School of Public Policy), National University of Singapore, in 1993–94. From 1994 to July 2005, he was a prisoner of conscience in Myanmar for 'seditious writings' and human rights work. He is at present working on the care and treatment of, and facilitating community support for, people with HIV/AIDS. A participant in Dialogue for Interfaith Cooperation and Peace-Building, his recent publications are: *Reality Check for Sanctions* (Hiroshima Peace Institute, Japan), *Poverty, Isolation and AIDS* (European Institute for Asian Studies, Brussels) and 'Transition in a time of siege: the pluralism of societal and political practices in the ward/village level in Myanmar/Burma', in *Active Citizens Under Political Wraps: experiences from Myanmar/Burma and Vietnam* (Heinrich Böll Foundation, 2006).

Editors' note

This publication uses 'Myanmar' in its title because that is the official name of the country, and is accepted as such by the United Nations. It has, moreover, been adopted increasingly in common usage inside the country, especially when speaking in Burmese. Its use in this publication does not represent a political statement of any kind. We, however, adopt the common practice of allowing authors in their own chapters to use whichever terminology they prefer for the country. With less well-known place names, where historical names have been used, we have added the current official name in parenthesis to avoid confusion.

Introduction

Monique Skidmore and Trevor Wilson

In the early years of the millennium, Burma/Myanmar endured several major crises that only aggravated the overall stress and the trying circumstances in which the country and the people found themselves. First, a banking crisis in 2002–03 brought the cash economy close to the point of collapse, from which it has still not fully recovered. Second, in May 2003 there was a serious political challenge to the military regime by the leader of the democratic opposition, Aung San Suu Kyi, to which the regime responded with characteristic ruthlessness in what has become known as the Depayin Massacre. Third, in late 2005, the regime peremptorily changed the official capital and uprooted the government and civil service from Rangoon to the new, isolated and still unfinished site of Naypyitaw.

Since the dramatic October 2004 leadership changes in Burma/ Myanmar, there has been little movement in the political situation. The government has essentially been on the defensive, nominally adhering to previous policies, while pursuing its purge against the military intelligence apparatus headed by ousted Prime Minister General Khin Nyunt. The National Convention that had reconvened in May 2004 resumed in February 2005 and has continued since then, as promised by the government before its adjournment on 29 December 2006 but still without representatives from the National League for Democracy (whose leaders remain in detention), from the second largest opposition party, the Shan Nationalities League for Democracy (whose leaders have been charged with high treason), or from the Karen National Union (with whom a cease-fire agreement has never been finalised and whom the government is fighting more vigorously than ever on the battlefield).

Meanwhile, the economy remains moribund, with investment and tourism staggering along at low levels. Western sanctions and informal campaigns against foreign investment have made small economic inroads, and living standards and disposable incomes have declined as prices climb and the value of the domestic currency falls. Evidence of any readiness to embrace economic reforms, even of the kinds adopted successfully by China or Vietnam, is lacking, and the prospects for effective engagement with the government about the options for changes in economic policy seem to be more remote than ever.

The education and health sectors continue to suffer from lack of government funding, and international assistance is insufficient to make up the difference. Standards of public health and education have declined disastrously. Meanwhile, the rule of law is honoured mainly in the breach, and widespread human rights abuses continue to be reported, but with less access than ever for independent outside monitoring of the human rights situation (especially with the refusal to allow the International Committee of the Red Cross to continue the full range of its operations independently).

The military regime managed to retain its tight grip on the country through these crises, but only by strengthening many of its repressive controls over the people, society and the economy. While the outside observer might be amazed that this could be so, for close observers and for the people of Myanmar themselves, it unfortunately comes as little surprise. Although the regime seems oblivious to international opinion and any criticism of its actions, it is struggling to make adjustments to its political rule through its so-called 'road-map' towards national reconciliation, and through its opaque attempts to develop and privatise the economy.

The contributions to this book were presented at the seventh Myanmar/Burma Update conference held in Singapore in July 2006 under the joint sponsorship of the Australian National University in Canberra and the Institute of Southeast Asian Studies (ISEAS) in Singapore. Along with Dr Tin Maung Maung Than of the ISEAS, the editors of this publication were co-conveners of the conference. The chapters represent an attempt by some of the world's most knowledgeable scholars of Myanmar/Burma to assess the political, economic, social and military situations as they stood in mid 2006.

In a new initiative, this conference also set out to examine some of the consequences of such a long period of authoritarian rule in Myanmar/Burma, this time looking at the impact on the natural and physical environment. Concerns are increasingly being expressed about the cumulative effect of years of neglect of Myanmar's natural resource endowment and its natural environment. The endangerment of Burma's mangrove ecosystem, the environmental, economic and social effects of logging, natural resource and wildlife depletion, and energy and pollution issues are examples of serious national problems that will have significant and lasting consequences for Myanmar/Burma in the twenty-first century. The contributors to the current volume are well aware of these issues and, after a broad consideration of the challenges for environmental governance, the contributions include some interesting case studies, all based on extensive in-country research into the reality of environmental management in Myanmar/Burma. While they do not necessarily seek to prescribe solutions, they illustrate dramatically why far greater attention needs to be paid to environmental protection and the sustainable aspects of development by the authorities as well as international donors.

The editors are grateful to the authors for their contributions, to Dr Tin Maung Maung Than and his colleagues in Singapore, who started the process of compiling the papers from the July 2006 conference, and to Asia Pacific Press for its support in publishing this collection.

Monique Skidmore and Trevor Wilson
Canberra
January 2007

1 The political situation in Myanmar

Vicky Bowman

The State Peace and Development Council (SPDC) is the most important actor in Myanmar's political economy.

This chapter focuses on the political situation in Myanmar in mid 2006 through the prism of the implementation of the seven-step 'road-map' of the SPDC, announced in August 2003 (Table 1.1). Outwardly, the implementation of this road-map appears glacial, with three years already devoted to step one (the resumption and completion of the National Convention to draw up a new draft constitution). But the road-map provides a framework that can be used to consider the wider political situation, as well as the SPDC's agenda and activities—declared and undeclared—and the responses of the opposition and the prospects for the future. The wider aspects of the road-map implementation can be considered to extend to the continuing war of attrition against Aung San Suu Kyi and her party, the National League for Democracy (NLD), in addition to other opposition elements which have strong name recognition, such as members of the 1988 student generation, and the SPDC's attempts to eliminate or suborn all armed opposition groups.

Table 1.1   The SPDC's seven-step road-map of 30 August 2003

---

1. Reconvening of the National Convention that has been adjourned since 1996

2. After the successful holding of the National Convention, step-by-step implementation of the process necessary for the emergence of a genuine and disciplined democratic system

3. Drafting of a new constitution in accordance with detailed basic principles laid down by the National Convention

4. Adoption of the constitution through a national referendum.

5. Holding of free and fair elections for *pyithu hluttaws* [legislative bodies] according to the new constitution.

6. Convening of *hluttaws* [assemblies] attended by *hluttaw* members in accordance with the new constitution.

7. Building a modern, developed and democratic nation by the state leaders elected by the *hluttaw*, and the government and other central organs formed by the *hluttaw*.

Source: *New Light of Myanmar*, 31 August 2003.

In parallel, the SPDC is trying to organise a political and administrative structure that can pursue its agenda during the latter stages of the road-map, a structure headed by the Union Solidarity and Development Association (USDA), which is being groomed to be the dominant political actor in a future multi-party state.

The SPDC has been trying to improve its popularity among the people, through enhanced publicity for its state-building activities and an anti-corruption drive among civil servants. This latter initiative, however—together with attempts to raise revenue by clamping down on tax evasion, the sudden move of the administrative capital to Naypyitaw and a lack of transparent, predictable or sound economic policies—is currently further slowing the nation's economy.

This chapter does not go into wider questions of Myanmar's history, or the present geopolitical situation, including the interests, policies and influence—or lack of them—of neighbouring countries and the wider international community, although these points need to be borne in mind when considering why the SPDC has adopted its current strategy.

The National Convention

SPDC Secretary One and National Convention Convening Commission Chairman, Lieutenant-General Thein Sein, had announced that the National Convention, adjourned on 31 January 2006 after three sessions since May 2004, would reconvene in the second week of October 2006.[1] He had previously noted that 15 chapters had been set down of the draft constitution, comprising some 75 per cent of the work (Table 1.2). This includes the controversial principles guaranteeing military participation in the Parliament (25 per cent of seats in the national, and 33 per cent in the regional, assemblies reserved for serving military) and their domination of key positions in the Executive.[2]

The October 2006 session would adopt the 'detailed basic principles' for the chapters tabled by the SPDC in early 2006, including relationships between *hluttaws* (or assemblies), rights and responsibilities of citizens and the role of the *Tatmadaw* (military). Judging by the process in previous sessions, once the convention reassembled, the proposals for these chapters would be adopted by a majority (but without a vote), in much the same form that they were tabled by the SPDC, although cosmetic changes could be included. A majority is easy to obtain since of the more than 1,000 delegates in the eight delegate groups, less than 100 were not hand-picked or vetted by the SPDC. Most elected political representatives, including those from the NLD, have declined to attend, since their leadership remains in detention and their offices outside Rangoon (Yangon) are closed.

Since the May–July 2004 session of the convention, few of those participating have bothered to engage with the process and make proposals for change. During that session, members of the 'Group of Eight', comprising ethnic cease-fire groups and 'other invited guests',

Table 1.2   Chapters of the Draft Constitution

'Laid down'	Remaining to be tabled by the SPDC (as of July 2006)
State fundamental principles	Election
State structure	Political parties
Head of State	Provision on the state of emergency
Legislature	Amendment of the Constitution
Executive	State flag, Seal, National Anthem and Capital
Judiciary	Transitory provisions
Tatmadaw [army]	General provisions
Citizens and their fundamental rights and responsibilites	

Source: *New Light of Myanmar*, 31 August 2003:1.

tabled significant—albeit poorly presented—proposals concerning the distribution of legislative powers between the centre and the regions. These were overruled by the SPDC, and the cease-fire groups now attend only because they are likely to face further pressure if they do not show up.

For all participants at the National Convention, whether hand-picked or otherwise, their chief objective is that it should be completed as quickly as possible. It appears that the SPDC is conscious of this, and of the cost of feeding, housing and entertaining more than 1,000 delegates, and is therefore accelerating the discussions by tabling a number of chapters simultaneously. It has not gone as far as committing to finish the process by a particular date.

It is possible, however, that the October 2006 session could be the last, particularly as many of the remaining chapters are fairly simple: for example, they deal with the state flag, seal, national anthem and capital. The basic principles set down as long ago as 1993 have already set the framework for some of the remaining chapters, although not the all-important provisions for amendment of the constitution. For example, the chapter concerning 'general provisions' will cover the designation of 'Myanmar' as the official language, and the establishment of a Constitutional Tribunal to interpret provisions of the State Constitution, to scrutinise whether or not laws enacted by the Union assembly, Region assemblies and State assemblies and functions of executive authorities of the Union, regions, states and self-administered areas are in conformity with the State Constitution, to decide on disputes in connection with the State Constitution between the Union and the regions, between the Union and states, between regions and states, among regions, among states, and between regions or states and self-administered areas and among self-administered areas themselves [and] to perform other duties prescribed in the State Constitution.[3]

The next steps on the road-map

The SPDC has consistently refused to provide a timetable for the next stages in the road-map process, much to the frustration of the international community, including the Association of Southeast Asian Nations (ASEAN), and appears to be keeping its options open by using the National Convention to provide it with flexibility, including over the timing of elections. There are some pointers that indicate that it is working to an internal timetable.

In November 2005, when SPDC representatives announced to embassies resident in Rangoon that the government would be shifting its administrative capital to a new site at Pyinmana (now called Naypyitaw), they informed the diplomats that at the end of 2007, plots of land would be allocated to missions on which they would be able to build new embassies. At the time, SPDC representatives were reluctant to allow foreigners to visit the area. By mid 2006, however, most official meetings with ministries were taking place in Naypyitaw, and the SPDC was keen to portray the new site as a pleasant and functioning administrative capital, to which it might have been expected it would be keen to encourage embassies to move (taking into account that most such moves take years). The most likely explanation for providing a target date for the official notification of the move of embassies two years hence could therefore be that late 2007 was expected to be the date when a new constitution, including the chapter designating Naypyitaw as the new capital, would have been adopted by referendum (road-map step four). This would thereby allow an official notification to embassies in line with diplomatic conventions.

Furthermore, some senior members of the government had indicated privately (with a certain air of desperation) that 'it will all be different after 2008'.[4] This suggests that 2008 is the year envisaged for elections of a semi-civilian parliament and assemblies (road-map steps five and six) after which the SPDC presumably hopes that Myanmar's relationship with its neighbours, and even the West, will be more normal. A normally well-informed Chinese diplomat also predicted as long ago as 2004 (at a time when the general view was that the SPDC was working to a 2006 timetable dictated by the forthcoming ASEAN presidency) that 2008 was a more likely internal deadline for a transition.[5]

Current political activity by the SPDC

In the meantime, the SPDC appears to be working on the intervening steps two and three of the road-map (see Table 1.1). The Attorney-General's department is thought to already have an almost-complete draft constitution reflecting the principles so far set down and those chapters to come, requiring little adjustment for the completion of step three.

Some had hoped that this and step two ('step-by-step implementation of the process necessary for the emergence of a genuine and disciplined democratic system') (Table 1.1) could have offered a space for a mechanism of national reconciliation involving opposition/civil society and the military/SPDC. But it appears that the SPDC is, instead, using the current period to try to garner support for its development activities, particularly among the rural population, while marginalising and eliminating all organised opposition. It is also engaging in *ad hoc* attempts to disarm (with negotiation) the smaller ethnic armed groups participating in cease-fires, in some cases rearming them as militias. This reflects the provision in the draft constitution that there will be only one *Tatmadaw* and that all those bearing arms in the country must be subordinate to it. Larger armed groups such as the Mon and Kachin expect that similar tactics will eventually be applied to them.

Indeed, stability, a single force, army unity, opposition to outside influence and a step-by-step approach to transition are the guiding principles of the SPDC's current approach, which is driven by an exaggerated fear of external interference in Myanmar, including a possible invasion by the United States and a deep-seated distrust of the NLD, Aung San Suu Kyi and all non-Burman groups. In the eyes of the SPDC, from its Chairman, Senior General Than Shwe, down, the aim of those opposing the SPDC, including the NLD, is to undermine the National Convention and revert to the 1990 election results and/or win the next elections with foreign assistance. All ethnic groups are regarded as wanting separation and independence, or at least federation—a dirty word for the military—and therefore should be treated with a firm hand militarily.

Than Shwe has also instructed his government to focus on 'union spirit' and avoid manifestations of regional or ethnic diversity.[6] This reflects his tactic of responding to ethnic nationality demands by broadly ignoring or over-riding them, rather than seeking imaginative solutions that could address the concerns of the ethnic nationalities about preserving their languages and culture within the SPDC's fundamental opposition to federalism.

The growing role of the Union Solidarity and Development Association

Another indication of the SPDC's apparent plans to move into the home stretch of the road-map is the enhanced focus on boosting numbers in and activities of the Union Solidarity and Development Association (USDA), and to a lesser extent the Myanmar Women's Affairs Federation (MWAF). The USDA is officially not a political party, but a social organisation. Since its formation in 1993, however, and particularly in recent years—although there was a brief hiatus after the USDA-orchestrated attack on Aung San Suu Kyi's convoy at Depayin in May 2003, when the USDA fell off the radar—the SPDC has been pursuing an internal strategy intended to make it the post-SPDC civilian political vehicle. The USDA is also taking a greater informal role in local administration, including a 'neighbourhood snoop' role (reinforcing the dislike that most of the population has for USDA cadres). The military is instructed to work in close cooperation with the USDA on irrigation, agriculture, economic issues and transportation in the regions, and central instructions have been for them to be present at all opening ceremonies of dams and bridges and so on, wearing USDA uniforms.

If the SPDC intends that the USDA should contest the election as a political party with its current name, it remains to be seen how it will overcome the self-created obstacles in the draft constitution that '[s]tate service personnel shall be free from party politics' (since most civil servants are required to be USDA members). [7] Since the strategic intention is clear, doubtless a solution will be found to fudge this and the fact that most USDA offices are on government property.

Courting popularity

Conscious of the *Tatmadaw*'s lack of support among the general population due to the demands made on them by the military, which shows up *inter alia* in recruitment problems and desertion, regional commanders have been instructed to improve discipline and morale among their forces, and to reduce the number of problems with the local population, including by minimising demands for forced labour, red-carpet welcomes and directives to grow crops against the farmers' wishes.[8]

After the huge increases in civil service salaries in April 2006 (another attempt to court support among a significant number of the population), the authorities instigated a clamp-down on civil service corruption on the grounds that this was no longer justified. The SPDC believes that since the main daily complaints of the 'Man on the Okkalapa Omnibus' relate to corrupt government officials and red tape, addressing this will improve its popularity. Officials in the trade, customs and tax departments have been arrested and reassigned, with heavy jail sentences handed out to officials, and disciplinary action has been taken against those government teachers who teach mainly outside school hours to supplement their low salaries. But with inflation wiping out most of the salary hike, any improvements are likely to be transient, particularly if they are not accompanied by simplification of the bureaucracy to eliminate the opportunities for graft, and a reorientation of civil servants towards serving the public rather than the military leadership.[9] Also, a number of well-known government figures and their wives appear to be untouchable, which undermines the credibility of any anti-corruption drive. Like most cultures, the Burmese have an adage equivalent to 'a building leaks from the roof'.[10]

The SPDC, and Senior General Than Shwe in particular, appear to be focusing on building support among the rural population, which makes up 70 per cent of the country, in the belief that they are more straightforward and honest and less likely to support opposition politicians or align themselves with urban intellectuals.[11] (That said, recent high-profile attempts to improve electricity supply by doubling the number of Electric Power Ministers suggests that the SPDC remains concerned about the urban population's anger about regular electricity blackouts). The senior leadership has instructed ministers to bombard the state-run media with facts and figures about infrastructure, in the belief that if the population is aware of the number of roads, hospitals and bridges built since 1988, they will support the SPDC and, by extension, the USDA. As part of this public relations drive, Information Minister, Brigadier-General Kyaw Hsan, has revived regular press conferences and has taken to bribing or forcing the non-state media into running coverage favourable to the SPDC. He has also increased attacks on the vernacular radio stations beyond his control (the BBC, Radio Free Asia, Voice of America and the Democratic Voice of Burma), on whom the majority of the population relies for domestic news.

The possibility that placing fewer demands on the local population or providing them with information about roads and bridges will lead to more favourable views of the SPDC is slight. The SPDC's approach is undermined, not just by critical radio stories, but by a shared common experience among most citizens of bad local governance and abusive local military-run administrations. It is also not helped by the continuing campaign to carpet the country with seven million acres of 'physic nut' (the castor-oil plant, a source of bio-diesel) to promote fuel self-sufficiency. This centrally directed project contradicts any directives to win the hearts and minds of farmers. Throughout 2006, the campaign received daily, blanket coverage in the state media as each of the commanders in the 14 states and divisions competed to show how they were meeting their 500,000-acre target. Even if it makes sense to develop some alternative energy supplies, the fanaticism with which the SPDC is approaching the planting of physic nut is regarded by the general populace as, at best, a perverse superstition and, at worst—by those who are forced to buy or plant the trees, or lose their land to plantations—a further abuse of their freedom and livelihoods.

Marking enemies

As part of its media campaign since 2005, the SPDC has intensified its public attacks on anyone it perceives as a possible political challenge, such as the NLD and the '1988 students'. The number of articles in the state-run media seeking to discredit the NLD as 'Western stooges' and 'axe-handles' and the verbal attacks on individuals increased in frequency and rancour. The SPDC's political approach towards those who dare to disagree with it was to identify them as enemies, and this intensified after the ousting of Khin Nyunt in 2004. In the filing cabinets of the military, the category of 'enemy'/'potential enemy' is a bulky one, encompassing well-known political figures such as Aung San Suu Kyi, former student leader Min Ko Naing, all NLD members, non-Burman ethnic groups (in particular the Shan) and, above all, the Shan State Army (South), Muslims, businessmen and former members of Khin Nyunt's Military Intelligence and his supporters. Indeed, it sometimes seems that, in principle, anyone outside the military should be considered an enemy. This includes foreign governments who are privately labelled enemies, even those such as China and India who publicly avoid criticism of the SPDC. For the SPDC, such governments could be considered temporary allies, but should always be treated with deep suspicion (something that has rendered attempts by countries to engage with the SPDC a frustrating experience).

Meanwhile, domestic enemies continued to be vilified, locked up, harassed and excluded from economic opportunities, or attacked through military means, in the case of the armed groups. While Senior General Than Shwe could have a personal and deep-seated antipathy towards Aung San Suu Kyi, dislike of her runs deep within the military, reinforced by almost two decades of indoctrination, as does the mistrust of the other categories of political opponents. Unfortunately, this is mirrored by an equally deep-seated mistrust of the military (and/or Burmese) among many of those categorised as enemies, and in particular those ethnic minorities who have borne the brunt of the past four decades of conflict.

The opposition

'Organised' opposition, whether in the form of the NLD, the 1988 students or ethnic groups, remained weak, harassed, divided and suffering from lack of effective leadership and experience, including in how to approach negotiations and build consensus. Their main objective is survival, as parties, groups or individuals. They have failed over the decades to come up with ideas that might have awakened the interest of the SPDC leadership in working with them, by addressing their key concerns, such as a continued role for the military, or their personal security. Yet their constant focus on the past, including the 1990 elections, rather than on the SPDC's road-map agenda, has further entrenched the SPDC view that there is no point in dealing with them.

Having marked them indelibly as part of a Western conspiracy, the SPDC has now clearly decided that marginalisation of the NLD is feasible and effective. It is not clear whether the party will ultimately be deregistered, but the threat has been made. NLD members in the districts are being systematically forced to resign and publicly criticise the party or face harassment in their daily lives, and even imprisonment on trumped-up charges. Many erstwhile activists are focused on personal, charitable or business concerns. Others, and the wider public, avoid contact with politically active groups, since these are punished by an SPDC jealous of the attention given to key opposition activists. The majority of the population, while privately opposed to continued military rule, remains focused only on the daily struggle to survive.

Meanwhile, the uneasy truce with the Karen National Union (KNU) has been put under pressure by increased fighting between the SPDC and the KNU's second and third brigades in the Toungoo area, and widespread human rights abuses against civilians forced to flee the fighting. Major operations are likely to continue against the Shan State Army (South). Other ethnic armed groups with cease-fire agreements with the SPDC are under increasing pressure to disarm, and their economic and political activities are being constrained if they do not do so. There is no sign that the SPDC and the ethnic groups will be able to bridge the gap between the latter's call for federalism and the former's abhorrence of it.

Prospects for a referendum and elections

In the SPDC senior leadership's mind, their political strategies to strengthen organisations supposedly loyal to the army are bearing fruit. They regularly 'count their votes', basing them on estimates of membership of the MWAF and USDA (currently at about 22 million, out of a national population of about 50 million, and rising, boosted by various incentives, such as the right to pedal a trishaw late at night).[12] As a result, the leadership is reportedly increasingly confident of securing its own future, and of therefore pushing ahead with the final steps of the road-map.[13]

Although an election could take place as soon as late 2007 or early 2008, there has so far been no sign of any preparation to run a referendum or election according to international standards. In particular, no preparation appears to have been made to update voter lists, which should include not only those attaining the age of 18 since 1990, but those who have never been registered by the central government, the majority of whom live in remote or cease-fire areas.

The cease-fire groups have not facilitated the prospects for this, having resisted for many years the adoption of registration mechanisms recognised by Yangon. One government official commented that this issue would have to await the referendum and new constitution.[14] (This raises the question of whether unregistered citizens would be disenfranchised from the referendum itself.) In 2004–05, there were rumours of preparations for an imminent census, which might have been a precursor to establishing a new voter register. These rumours have, however, stopped.

An election requires a significant investment to meet international standards for voter registration, civic education, provision of transparent ballot boxes and other things if it is to have any chance of being considered genuinely free and fair (as attested by the millions spent by the international community on post-conflict elections in Congo, Afghanistan, East Timor, Iraq and elsewhere). In its present cash-strapped state, the SPDC is unlikely to be able to make the necessary investment, even if it were in its interests to have a free and fair election. But it will also be unwilling to see any international involvement or observation, even if it brings with it funds to run the election, since it will perceive this as interference. It is likely to run a shoestring operation, with the laces carefully tied. According to one government official, the leadership has reviewed the way in which the 1973 referendum on the 1974 constitution was conducted, with separate 'Yes' and 'No' boxes (white and black respectively), the latter requiring a long walk to reach. Despite their supposed confidence that they can now carry the rural population with them, it is widely believed that they will take whatever measures are necessary to avoid the 'mistakes' of the 1990 election, which produced a landslide victory for the NLD. There is even speculation that the SPDC could skip a full plebiscite on the draft constitution (step four of the road-map) and simply opt for a nation-wide mass rally, citing the support of 22 million USDA members as proof that the referendum has majority support, similar to the manner in which they have run the National Convention.

While the 1973 referendum was marked by relatively high levels of participation and interest (although the official turn-out figures—more than 90 per cent—were likely exaggerated), any future referendum and elections are likely to see a low real turn-out. This will reflect partly problems of registration, but a major factor will be voter apathy, a lack of interest in politics—growing among the urban young—and the nature of the draft constitution, which few believe will make any significant difference to their lives. Indeed, the lack of public and private debate on any of the steps of the road-map, including the National Convention, constitution and elections, is striking. Apathy is likely to favour the SPDC.

None of the groups constituting an organised opposition (the NLD or the larger cease-fire groups) had indicated their approach towards either a referendum or election, including whether they would opt to participate in elections, if they were able to do so. They understandably prefer to wait to see how the SPDC approaches an election. They would also be aware of the provisions in the draft constitution that disqualify from election to the *Hluttaw* 'a person who commits or abets or [a] member of an organisation that commits or abets acts of inciting, making speeches or issuing declarations to vote or not to vote'.[15]

Poor prospects for progress

Although things could be 'different after 2008', there were no indications in the middle of 2006 that current changes would result in a fundamental shift either in the way Myanmar was governed, or in its relations with the international community. The present political situation in Myanmar, therefore, offers gloomy prospects.

In particular, there is currently no prospect of an end to *de facto* military rule in Myanmar, as codified in the much-contested sixth guiding principle of the National Convention/draft constitution requiring 'the *Tatmadaw* to be able to participate in the national political leadership role in the State'.[16] In other countries emerging from military rule, such as Indonesia and Thailand, timed phase-outs of constitutional military involvement in politics and government have been spelt out. But there is no sign of this in the draft Myanmar constitution. Although such a constitutional phase-out alone will not be enough to demilitarise the State, it at least provides a framework containing the ultimate prospect of civilian government. This would be something that the population and the international community could look forward to and that might help to lift the gloom.

It seems likely, however, that if current political trends continue, any elections held under the road-map will not come close to meeting international standards for a free and fair poll. Although the detailed basic principles of the constitution concerning elections and political parties are not yet available, let alone an election law that would be based on them, the SPDC's current approach towards the main legally registered political party, the NLD, suggests that in practice it would take all measures necessary to avoid a level playing field at the time of the election.

The consequence will be that the Myanmar/Burma 'brand' will continue to be associated internationally with human rights abuses and 'that woman'. The deadlock with the international community, and in particular the United States and the European Union, and international financial institutions such as the World Bank, the Asian Development

Bank and the International Monetary Fund, will continue. Furthermore, if the end of the road-map promises more of the same governance under different hats, there appears little likelihood of better informed or more accountable economic policies, or more transparent rule of law, which could attract foreign investment. As a result, Myanmar will not attract the international public or private investment it needs to benefit from its geographical situation and potential, and it will continue to be a weak link in the development of the Asian region.

More importantly, the outcome looks like failing to open the way towards a new era of politics for Myanmar, which might begin to resolve the tensions and inequities that precipitated the past five decades of internal conflicts, including the uprisings in 1988.

Notes

1 *New Light of Myanmar*, 3 September 2006:1.
2 *New Light of Myanmar*, 30 July 2006:16.
3 Available from http://www.ibiblio.org/obl/docs/104principles-NLMb.htm
4 Personal communications with several SPDC officials, mid 2006.
5 Personal communication, 2004.
6 Personal communication, 2005. This instruction probably lay behind the sudden order to the luxury hotel under construction near Putao to change its name from Lisu Lodge to the less ethnically identifiable Malikha Lodge; and the instruction to ban Mon students from wearing national dress to university every 'Mon-Day' (*Burmanet news*, Issue 3039, 2–5 September 2006).
7 http://www. ibiblio.org/obl/docs/DBP-LEGISLATURE.htm, Paras 33(j) and (k).
8 Personal communication, 2005.
9 As typified by the requirement for local education or health officials to waste a day waiting to greet a visiting military VIP rather than getting on with their jobs.
10 *Kaun-gá-sá mo: má-loun-hmá-táw*.
11 In *Behind the Teak Curtain* (2004), Ardeth Maung Thawnghmung explores the attitudes of the rural peasantry since independence and shows that they did—at least before 1988—tend to identify more strongly with the military, which has risen from rural stock, than distant urban élite politicians.

12 Network of Democracy and Development (2006) outlines some of the positive and negative incentives for USDA membership.
13 Personal communication with a senior official, June 2006.
14 Personal communication, late 2005.
15 http://www. ibiblio.org/obl/docs/DBP-LEGISLATURE.htm, Para.33(h).
16 *New Light of Myanmar*, 31 August, 2003:1.

References

Thawnghmung, A. M., 2004. *Behind the Teak Curtain: authoritarianism, agricultural policies and political legitimacy in rural Burma*, Kegan Paul International, London, New York.

Network of Democracy and Development, 2006. 'The Whiteshirts: how the USDA will become the new face of Burma's dictatorship', Network of Democracy and Development, May. Available from http://www. ibiblio.org/obl/docs3/USDAFinal1.pdf

Acknowledgments

The views expressed in this article are the personal views of the author and do not represent the view of the British government.

2   A Burmese perspective on prospects for progress

Khin Zaw Win

It has been said that history is written by the victors. In the same vein, progress can be said to be defined by who, or which side, is carrying it out. Superficially, a linear motion forwards is assumed, just as for the march of modernisation or of civilisation itself. But in reality, the very term 'progress' is a loaded one, and that it stands for a complex process.

It is not surprising, therefore, that attempting to define progress in Myanmar's case is a precarious undertaking. The incumbent regime steadfastly maintains that immense strides have been made since it assumed power in 1988, which is not untrue in a number of aspects. On the flip side—something that the opposition never tires of pointing out—there are other facets that are not causes for celebration. Any number of individuals and organisations base their assessments on the simple formula that lack of democracy equals lack of progress. In the face of such daunting circumstances and the complexity of the task, one has to pin down what could confidently be labelled progress. At the same time, instead of arguing about what it means or represents, a more productive endeavour would be to identify turning points—those that have passed and those that could come—that could, if handled properly, make a difference for the country.

The argument about progress is especially intense when it comes to the question of which direction and through whom such progress is to be made. The general direction indicated by the events of 1988 and their immediate aftermath was towards a democratic system and a market economy: in short, overturning the straitjacket system established by former Head of State Ne Win and his Burma Socialist Program Party (BSPP). But beyond that initial, universally agreed step, differences yawned wide. The opposition called for a speedy transition to democracy and, after the 1990 elections, an immediate transfer of power to an elected government—which meant a National League for Democracy (NLD) government. And ever since, whatever hopes there might have been for political progress were effectively sidelined by the running battle over the 'transition to a democratic government', a bitter wrangle over an endless series of issues. By extension, advances and improvements in other vital sectors are being held up.

It is especially ironic for a country as badly in need of development as Myanmar to find itself blocked by a political process proclaiming its devotion to the betterment of the country. The change over to a democratic system espoused by the main opposition is very much along the lines of the liberal consensus model. It should be clear by now that this has run into two main difficulties.

Firstly, there is increasing evidence in developing countries that rises in per capita income precede the emergence of democracy and not the other way around; and that good governance, far from being a precondition for rapid growth, is typically an outcome of successful economic development (Khan 2004). There is a convincing case for poverty and lack of economic development being the context—together with internal conflict—in which poor leadership played itself out in Myanmar's earlier experiment with democracy. Developing country status conferred in late 1987, to which is added the destruction and dislocation of 1988, does not augur well for the prospects of a quick jump to democracy. Amazingly enough, precisely such an opportunity opened up in September 1988 when the BSPP government was on its last legs, only to be passed over by the leadership of the democracy movement. There have been others in subsequent years, but none of these were taken advantage of. The prospects themselves were there—as they are now—but turning them into reality was another matter.

This brings us to the second difficulty, one that is directly related to the NLD and stems from the way it has played its cards. Besides the precariousness of the preconditions for democracy, the NLD's strategy— or lack of one—has practically written off the tenability of the liberal consensus approach to building democracy in Myanmar. This has come about because one individual's wishes have been allowed to fashion an entire action plan for democracy. Moreover, this is a plan that, to force through a political settlement, does not hesitate to call for economic measures to be taken against a country struggling to develop itself.

The 18-year political tussle in Myanmar between the military government under the State Peace and Development Council (SPDC) and the democratic opposition has undoubtedly dragged on for far too long, and there are indications that, by the very fact of its lengthy unproductiveness, it has made itself almost irrelevant to the majority of the population. The much-touted and long-proposed solution from the main opposition party has itself become a problem. The times now cry out for a 'second opinion' of what is really ailing Myanmar, and therefore for a fresh, more effective, line of treatment.

Meanwhile, the country continues to reel from the effects of a poorly managed transition from a centrally planned economy. This has been exacerbated by sanctions imposed by the West. For many people, the difficulties of daily existence have meant that political issues receive less attention and priority, particularly with the untenable nature of the present democratic alternative.

Against this backdrop, an assessment of prospects for progress could begin with peace agreements.

The winding down (which began in 1989) of the civil conflict is an event that is under-acknowledged and underrated. Besides that, it has not been well integrated with the concomitant democratic transition and, more importantly, post-conflict political needs have not been attended to. One could say that the promise of the 1947 Panglong Agreement—which many perceive to have been betrayed, and hence the cause of the decades-long ethnic minority rebellions—is being given renewed attention. The ethnic nationalities expect nothing less.

That the majority of the cease-fire agreements continue to hold can be seen as a vote of confidence in the present government, as well as attesting to the exhaustion (in all senses) of armed rebellion as a mode of political action. Nevertheless, the peace agreements have not been followed up satisfactorily. Moreover, the democratic opposition has been equivocal about them. The prolongation of the political stand-off has meant that pockets of armed conflict persist in the country, with all the negative consequences that attend in their wake. The final cessation of conflict should have occurred earlier, were it not for the absence of a settlement among the main actors in the capital.

It has been a lost decade for democracy in Myanmar, and needlessly so. Democratisation is going to take longer and the Third Wave in Myanmar is being drawn out. Does that mean that development is to be drawn out, too? And would not devolution of governance be needed as much as or even more than democracy, because minorities and smaller parties have little to gain from majoritarianism in whatever garb it arrives? Consociational democracy,[1] which has been adopted successfully in other countries in the region, is still some way off. The new constitution being drawn up now, controversial as it is, does provide for 14 sub-national administrations and legislatures as well as six smaller bodies for autonomous regions.

When centre–periphery and majority–minority relations in Myanmar are considered, the way that *Bamar* or Burman (majority) politics has impacted on minorities is a crucial but little studied issue. Now for the first time, prevailing realities are bringing about a system that is not wholly majority centred.

No matter how the new constitution is viewed, there is no denying that a good deal of decentralisation and devolution of power is going to take place, and this could precede democratisation. In the present context, devolution would appear to promise more pluralism and ultimately a sounder democracy in the future. In other words, the greater exercise of democracy at the provincial level would pave the way for eventual full liberalisation at the national level. In Myanmar's case, this would be a sound and workable formula.

The unavoidable issue, then, is to ensure that the ethnic groups are equal to the task. There is a need for people in the states to build up their capacity to participate in governance. The standard and oft-proclaimed statement that democracy will solve everything does not apply here—or anywhere else for that matter. Stepped-up development in the ethnic states—including the cease-fire areas—has been going on for more than a decade. This part of the peace dividend could have been more fruitful were it not for the throttling of development assistance and investment from abroad.

The main democratic opposition: a penchant for losing opportunities

The main opposition continues with its fixation on elections, particularly the 1990 elections, which were but one event in the transition. By doing so, it is destroying its own credibility, and the difficulties this causes have spilled over to the democratic movement as a whole. Other than those espousing the extreme hard line in the regime, no one would like to see the main opposition organisation totally excluded. Despite rumours to the contrary, the NLD is being allowed to continue as a legal political party. It has a place in Burmese politics, no doubt, but that place is going to be different from what the organisation imagines it is entitled to. If it refuses yet again to face up to reality, it would be assuring its own extinction. It will be undergoing an involuntary make over; and its new niche will be the product of a process of triangulation effected by the military, the country's situation and its own capability. It is like water finding its own level.

There have been at least four notable, '24-carat' opportunities for the opposition leadership since 1988. These were not flashes in the pan; they were solid, serious openings that stretched on for months, if not years.

The first was from January to May of 1989. From about the time of the death of Daw Khin Kyi, General Aung San's widow, the then chairman of the ruling military council, General Saw Maung, made a sustained conciliatory overture to the democratic forces (which were very strong at that time). His gesture was aimed particularly at Aung San Suu Kyi.

After Aung San Suu Kyi's first period of house arrest and the ouster of General Saw Maung in April 1992, a second opening emerged, beginning with the removal of restrictions against Aung San Suu Kyi in July 1995. Not only was this opening thrown away, it was turned into a long period of skirmishes during which Aung San Suu Kyi attempted many variations on the confrontation theme and the military tried to contain her.

Particular mention should be made of the talks that the military council proposed to the leadership of the NLD (minus Aung San Suu Kyi) in September 1999. These talks were scuttled at the last minute when the NLD leadership followed Aung San Suu Kyi's request and refused to attend.

The third opportunity came after the failed coup attempt by ex-party chairman Ne Win's family in March 2002. More will be said on this below.

The final window opened between May 2002 and May 2003. Following on from the efforts of the United Nations through its Special Envoy and those of Western countries, restrictions on Aung San Suu Kyi were again lifted. General Than Shwe reportedly did this despite grave reservations on the part of Generals Maung Aye and Khin Nyunt.

Although it was not made public at the time, there had been meetings between Aung San Suu Kyi and General Than Shwe, together with other high-level talks with some members of the Cabinet. Unparalleled opportunities had in fact been provided. If anything resembling the NLD statement of 12 February 2006 had been issued at that juncture (that is, three years ago), Myanmar's political and economic fortunes could have been so much different. It is regrettable that Myanmar's democratic transition is going to take longer. The central cause of this delay is personal and individual.

Not only has the democratic leadership alienated the top echelons in the military, it has opened a wide gulf between it and the middle-ranking officer corps—a development with even greater implications for the future.

Beyond merely pushing an amicable settlement on the democratic transition out of reach, Aung San Suu Kyi's *konfrontasi* has precluded the intra-societal meeting of minds that is necessary to get a handle on the country's problems. The breakdown in inter-élite relations has led to a wider breakdown.

The armed forces: breaking with the past

With Senior General Than Shwe's assumption of the top military council post in 1992, there was a pronounced distancing from former strongman Ne Win's influence. Many people at the time believed that Ne Win still exercised control over the ruling establishment, particularly through the agency of his loyal protégé, Khin Nyunt. Any regime anywhere would have been hampered and threatened if it had to keep looking over its shoulder, so to speak, most of the time. In addition to the many pressures the regime was being subjected to, this dangerously unpredictable ex-dictator was lurking in the wings.

After Ne Win stepped down in 1988, members of his immediate family—his daughter, son-in-law and grandsons—were able to set up substantial business interests. They still had clout and could still throw their weight about, but eventually they came to realise that power was no longer their exclusive domain. Resentful of the new competition and of their interests gradually being relegated to the back burner, they plotted a coup to coincide with Armed Forces Day, on 27 March 2002. The authorities managed to nip this in the bud with a pre-emptive strike. Three military regional commanders, the chief of police and a number of lower-ranking military personnel were implicated and dismissed.

The SPDC's handling of this threat appeared to offer justification for two things (which are really a continuum): the rationale for its rule since 1988 was placed on still firmer ground and its argument for a continued political role was strengthened. Testimony of some eloquence was furnished for the regime's protestations that it was providing political stability and laying the ballast for the country's road to democracy. It is hard to imagine a democratic government handling a crisis like this on its own.

Then, too, a singular lack of discrimination and sophistication had become apparent on the side of the democrats. In realising where a common threat lay, and in its extirpation, an avenue for reconciliation and the reforging of a relationship had opened up. But, again, it was passed over.

Later that same year, two lieutenant-generals—Win Myint and Tin Hla—invested with a brace of important positions, were unceremoniously axed. The ostensible reason was corruption, with the military business entity Myanmar Economic Holdings involved. Win Myint, however, was linked to Ne Win, and it is quite likely that power maneouvres figured in his dismissal.

Of all the generals who since 1988 had made their pitch on the national stage and then exited, none had been as powerful as Khin Nyunt. Hand-picked by Ne Win in the mid 1980s to head the intelligence organisation, he had grown in power as well as ambition. If he had been content to make the positions he had achieved—intelligence chief, full general, prime minister—the pinnacle of his career, and planned to make way for a new generation, he would have enjoyed a statesman-like finale. It was quite plain, however, that he had his sights set on the presidency.

He had been steadily adding to and broadening his power base over the years. His strongest support outside intelligence and government circles came from certain of the ethnic paramilitaries and new business tycoons. His wife, too, had been the head of the two state-sponsored women's organisations. Khin Nyunt was behind some of the reforms that had taken place and there seemed the promise of more to come, but the wide categorisation of him and his camp as moderates, and the other bloc in the military as hard-liners, does not stand up. He did have the most international exposure of all the generals and thereby could not have missed the way the wind was blowing; on the other hand, he could have been carefully cultivating an image.

Notwithstanding the political identity make over that appears to have occurred, it should not be forgotten that the Defence Services Intelligence (DSI) under Khin Nyunt was one of the main props of successive authoritarian regimes. It was the major instrument of repression, particularly of the political opposition. One of its foremost tasks was the continuum of 'processing' democracy activists, from surveillance and arrest to interrogation, conviction and imprisonment. Over the decades, DSI had become the most powerful and feared organisation in the country and Khin Nyunt was its longest serving chief.

Even from the reasons publicised after his ouster, it can be gauged how intolerable he had become to the main echelons of the military. The accusations levelled against him included disobeying orders, corruption and failure to comply with regulations. The intelligence apparatus was described as being above the law and preying on the people. Yet, the DSI and Khin Nyunt were among Ne Win's more successful creations. Khin Nyunt's position as a Ne Win protégé and Western-trained intelligence officer—plus his ability—had brought him to the top. Gunning for the supreme state post, however, requires more than that. He knew this, and assiduously cultivated his image and popularity, but in the end things backfired. The dual contexts ultimately won over: the dissonance of the military intelligence chief in a professional, combat-hardened army that increasingly came to resent him; and heading the feared and hated secret police in a society in transition to a more liberal system.

The top echelons of the army are now going through another intricate and measured reordering and reconfiguration, even more so than before. The final line-up will take the State through the culmination of the National Convention, the drafting of the new constitution and the formation of the government under that constitution. An indication of who will fill the future top state posts is expected to become apparent as this process unfolds.

It used to be that the military regional commanders (who are major generals) were moved into the Cabinet or promoted together in one batch. Now, in a first instalment on 13 May 2006, of the 13 military regional commanders, two were given senior military positions vested with the rank of lieutenant-general, while three were given ministerial portfolios. Of the remaining eight commanders, one was appointed to the newly created Naypyitaw Command, so seven should have been awaiting their future assignments. The delicacy of this present rearranging is reinforced by reports (albeit unconfirmed) that deputy prime-ministerships that have been vacant since late 2002 are to be filled.

There are extreme elements within the establishment who would like to return to a neo-authoritarian system, but there are also other elements who oppose this and who are unhappy with unfettered authoritarianism. This is quite a positive sign. The extremes on both sides—within the establishment as well as the democrats—have to be opposed.

A new generation in the military is coming to the fore; it could have come even earlier were it not for certain tensions, intra-institutional as well as in the body politic. It is natural to have high expectations of every new generation and the present case is no different; however, when it comes to dealing with the present stamp of democratic politics, it would be unrealistic to expect a drastic change.

Opportunities forgone

It is noteworthy at this point that the conclusions from a detailed, long-term study of transitions from authoritarian rule in Latin America and Southern Europe echo uncannily the Myanmar experience.[2] In fact, they predate Myanmar's transition and, if they had been made known sufficiently, could have served as a guide. It is worthwhile to highlight the parallels, and even if the way to a successful, trouble-free progression was missed, at least it can reconcile Myanmar to its present predicament and offer some help.

The study's assertion that an active, militant and highly mobilised popular upsurge could be an efficacious instrument for bringing down a dictatorship, but could make subsequent democratic consolidation difficult and could provide a regression to more authoritarian rule, is remarkably prescient of Myanmar in 1988, to the extent of holding up a mirror to those events. It advises postponing for an undefined period the goal of an 'advanced democratic' transformation, pointing out the advantages and disadvantages of democratisation 'on an instalment plan'. It adds that by initially accepting the role of strong but loyal opposition, newly emergent parties could find that they have taken the best possible path to power, in terms of optimising their eventual electoral strength and minimising the immediate risk that they would be impeded from taking office by violence.

The final comments could give some indication of where Myanmar's political destination lies and offer some hope. It reminds us that the circumstances of the transition compel players to compete for space and pieces rather than struggling for the elimination of opposing players, and that political democracy is produced by stalemate and dissent rather than by prior unity and consensus (something that could come as a surprise to Burmese democrats). It emerges from the interdependence of conflicting interests and the diversity of discordant ideals, in a context that encourages strategic interaction among wary and weary actors. Transition towards democracy is by no means a linear or a rational process. There is simply too much uncertainty about capabilities and too much suspicion about intentions for that (O'Donnell 1986). It remains for the Burmese players to recognise this.

The democrats have missed tremendous chances in the past decade. There was a superb opportunity to establish a position—unheard of in 30 years—from which a political party and democratic leader could have positively influenced the Myanmar military institution, no mean task in itself. Furthermore, it could have laid the ground for the military—despite all the assorted hard-liners in it—to respect democracy and to respect a democratic civilian government. All this has now passed into the 'if only' realm.

It goes without saying that foremost among the hallmarks of good leadership is the ability to recognise and seize such opportunities. This, to Myanmar's detriment, did not happen. Instead what we witnessed was recourse to more of the same: an ingrained, even pathological fixation on drumming up a call to 'take to the streets' and to repeat the 1988 uprising. With Aung San Suu Kyi's continued dismissal of the overtures and opportunities proffered by the military council, it was inevitable that the SPDC would announce, as it did on 26 April 2006, that it would not negotiate with the NLD. Finally, common knowledge that the door has shut became officially declared policy. The expected closure of an option, however, signals the greater need to seek and work along other avenues.

For the polity as a whole, instead of an absolutist fixation on the individual, there is a need to work on systems, institutions, organisations and processes, all of which constitute the foundations of a viable and enduring democracy. When a heavily managed democracy is mentioned, it usually connotes perpetration by the military or other authoritarian system. But what if democracy comes to be heavily managed by a predominant democratic party? The continued invoking of the majority obtained in the 1990 elections could provide a platform in opposing the regime, but this majoritarian line also brings a hardening of views, an intolerance of diversity of opinions and the inability to cope with failings of leadership. What the country needs is a solution, not a litany of blame, demands and positions. A serious, committed search for answers is taking place, but regrettably not in the orthodox political organisations.

Like it or not, the case for 'tandem' governance has been strengthened. The elected and non-elected components of such a system can complement, balance, buttress and correct each other. The very fact of cohabitation spells the establishment of a relationship. At the very least, this could pre-empt the uncontrollable divergences that have plagued the country in the past.

The experiences of Thailand and Chile have shown that democratic consolidation is possible even under military-imposed authoritarian constitutions. In Chile, the authoritarian constitution was circumvented and put to work for a democratic purpose through an institutionalised party system, a tradition of rule of law and capable political leadership. The fact that these conditions are only minimally fulfilled in Myanmar could alone account for the poor showing of democratic consolidation.

The Chilean case study mentions that 'a minimum of democratic rights and institutions granted by the 1980 constitution have been skillfully used by able democratic agents to foster political competition, to a level that most scholars never thought possible'. The neo-institutionalist claims that only the removal of authoritarian institutions would clear the way for democratic consolidation seem overstated.

The political system instituted by the 1980 constitution could be called a sub-minimal democracy. The review process since then, especially after reforms in 1989, illustrates a case of slow, negotiated transition from authoritarian governments that have not suffered military or political defeat.

Another lesson from this case study is that pro-democracy constitutional reforms result more from changing political interests and depolarisation of actors than from much-cherished democratic principles or strong political will. A systematic view of institutional change is proposed, rather than a linear or mechanical approach more congenial to neo-institutionalism (Esteban Montes and Vial 2005).

In this regard, it would certainly not be out of place to recall one of the central issues during the struggle for independence half a century ago. After World War II, bowing to realities and to the Atlantic Charter, London offered dominion status to its colony. The majority of the *Bamar* population, however, caught up in nationalist fervour and almost totally behind the leading party, the Anti-Fascist People's Freedom League, turned the offer down and demanded total and immediate independence. Even membership in the Commonwealth was declined. In the prevailing ardour, hopes and assumptions fed on each other, leading to expectations that self-government would work wonders; and there was a childlike faith in the national leadership. What was left out of the reckoning was a grasp of the indigenous political culture and the nature of the political élites.

What ensued is common knowledge: all-out civil war in all its brutality, in a land that had been trying to rebuild from the ashes of World War II. There are consequences—political, military, ethnic and economic—that remain to this day. With sobering hindsight, quite a few people, even the communists, now privately admit that independence could have been postponed beneficially for five or so years.

Myanmar has been at the mercy of inept politics for far too long. The present contentious era has become the longest, most unproductive and damaging of all—so much so that political parties, leaders and processes have all lost credibility. The public comes to realise that politics as it is practised now is not providing the answers or a way forward. There is so much emphasis on democracy, while the extreme weakness in associative capital is disregarded. Because of this, parties, organisations and even the military have foundered repeatedly. The task ahead includes a salvage operation for the democratic movement as a whole.

With regard to the course that the present regime has embarked on, there could be scattered suggestions of citizens having to acquiesce or having to give up what is deemed to be an unequal struggle. But in the larger picture—the present national context—there really is no other way.

One prospect to bet on

There is an urgency in the quest for new forms of the State, development and political discourse that are harmonised with (or at least not discordant with) Myanmar's historical and sociological foundations. When one looks beyond appearances, beyond the struggle between creaky and recalcitrant authoritarianism and the brave forces of democracy and liberalism, there is on the one hand the phasing out of a party system that has been the bane of the country practically since its inception during the colonial period, letting down the people continuously. The democratic leadership serves only to hasten and even facilitate the passing of an unworkable scheme. On the other hand, there is a desperate attempt to fashion a substitute. The military regime running the country in the meantime is only utilitarian, even incidental; someone has to keep things going.

There is general agreement on the weakness of the State in Myanmar, something, moreover, that is coupled with a strong society. The prospects for strengthening the State do not seem bright for the foreseeable future, no matter how the military regime sees itself or tries to keep up appearances. Due to an unprecedented converging of circumstances, the shape of the future state structure as delineated in the principles laid down for the expected constitution includes 14 provincial (state and division) governments and legislatures as well as six smaller entities. For all its detractors, the new configuration will draw a great deal of attention and energy. Add to this the new institutions that have paradoxically emerged from the decades of internal conflict, along the lines that scholars such as Charles Tilly (1975) described. Internal conflict resulted in periodic administrative reforms, such as the great centralisation of the BSPP period. With its abject failure, a process of creative destruction can be said to be under way and a mix of centralisation and decentralisation is being propounded. In the present cycle, it is undeniable that incipient ethno-political entities have to be accommodated.

Leaving aside the fine print for the moment, Myanmar is on the brink of federalism by any other name, something much longed for and aspired to in many quarters. Barring major missteps, the scope has definitely opened up for sub-national states to grow directly out of regional societies, communities and ethnic groups.

There will be a military presence, and that of state-sponsored organisations such as the Union Solidarity and Development Association can be considered likely. But there is a good chance that the impending provincial establishments, particularly those in the ethnic regions, will be distanced from and even spared the dead hand of *Bamar* politics, and thereby will attain a level of viability and efficiency.

Conclusion

Bringing democracy and human rights to Myanmar cannot be equated with elevating one individual to power. Nor can it be equated with attempting to achieve political ends, however commendable in name, by means of a poverty-promotion program.

It is to be hoped that ultimately the armed forces and the democratic parties will come to realise that state power is not something that is up for grabs and that only a stable and liberal system can set and safeguard the crucial structures and processes that order such power. In other words, institutions have to come before any single individual. For the armed forces, after being plagued by uncertainty, tension and crisis stemming from within, a hard lesson being learned is that the best assurance of its long-term integrity and stability—particularly when it comes to leadership transitions—lies with stronger state institutionalisation.

For everyone concerned, but particularly for the 'other' democrats in Myanmar, the most important thing is not to let the present stalemate become a perpetual hindrance in the larger task of building a liberal democracy. The issue of a compromise constitution will have to be faced, initiating the unending series of compromises that amount to a democracy. One of the core tasks has to be the thrashing out of a form of cohabitation. Beyond even all that, what Burmese society is engaged in—if it would only realise it more fully—surpasses what either the military institution or any political party presumes itself to be the arbiter of. The processes will not be easy, nor will the outcomes satisfy everyone. The country is faced with the overarching issue of nation building.

Assumptions are not enough and one must look beyond the hype, as well as beyond the histrionics, the polemics, the posturing and even beyond the personal suffering. In the emotion-charged atmosphere after 1988, a majority threw its support behind what seemed to be the solution. Now it is clear that fate has placed Myanmar between two grey-hued organisations, both professing a path to democracy. In both, decisions come from the top: participatory decision making is unheard of. More and more, democracy has been relegated to the status of an excuse, a window-dressing, a stage prop and a handy rallying cry. The real issue—as anywhere else—is personalities vying for power. A real concern for democratisation means concern for the plight of the people; this, it should be pointed out, is missing.

When this is realised, the 'struggle' loses quite a bit of its meaning. What is happening now can be seen as only the inevitable working itself out. It is universally agreed that there should not be a return to an authoritarian system, but if the military should not continue with the systems, constructs and methods of the unhappy past, the same should apply for the other side—the democratic opposition.

In one sense, it is a huge disappointment. In another sense, it is a lesson learned and applied—in the nick of time. Indeed, the real business begins only now, when the dust has cleared and hot blood has cooled: the business of democratisation, that of building a unified yet decentralised nation, the bringing of the economy back on an even keel and the climbing back from developing-country status. These processes are indeed already under way. They could and should be helped along; no one who is even remotely concerned with Myanmar's future could afford to neglect them, much less hinder them. A country that has been relegated to developing-country status and has a half-century of travail behind it should not be penalised for following the most natural, realistic and feasible path before it under the circumstances. This is exactly what Myanmar has to do.

Notes

1   Consociationalism is a form of government involving group representation by élites, and is suggested for deeply divided societies. According to Rupert Taylor, 'Consociationalism advances a system of consensual multi-ethnic power sharing as opposed to majority rule.'
2   The detailed, long-term study is *Transitions from Authoritarian Rule: prospects for democracy in Latin America and Southern Europe* (O'Donnell and Schmitter 1986), a project of the Woodrow Wilson Center, 1979–81. It comprises four volumes of edited papers and a final volume of conclusions.

References

Callahan, M., 2003. *Making Enemies: war and state building in Burma*, Cornell University Press, Ithaca.

Esteban Montes, J. and Vial, T., 2005. *The Role of Constitution-Building Processes in Democratization*, International Institute for Democracy and Electoral Assistance, Stockholm.

Khan, M.H. 2004. 'State failure in developing countries and institutional reform strategies', paper presented at the *Annual World Bank Conference on Development Economics—Europe 2003*, World Bank, Washington, DC.

O'Donnell, G.A. Schmitter, P.C. and Whitehead, L., 1986. *Transitions from Authoritarian Rule: prospects for democracy in Latin America and Southern Europe*, The Johns Hopkins University Press, Baltimore and London.

Tilly, C., 1975. 'Reflections on the history of European state-making', in C. Tilly (ed), *The Formation of National States in Western Europe*, Princeton University Press, Princeton.

3   Of *kyay-zu* and *kyet-su*: the military in 2006

Mary Callahan

To many observers, the *Tatmadaw* (Burmese, for 'armed forces') of 2006 appeared omnipotent. Its senior officers ran a state that had eliminated or neutralised major rivals. It had concluded truces with, or obtained surrenders from, nearly all of its former armed adversaries. The military completed a breakneck-paced expansion of its personnel, from 180,000 in 1988 to close to 400,000 in the mid 1990s (probably about 300,000 today). Garrisons now dot the map of the whole country, a vast change from the pre-1988, post-colonial setting. Commercial enterprises associated with the armed forces play significant roles in many sectors of the economy and individual military units run an extensive array of industries, plantations, road check-points and other revenue-raising ventures.

But the *Tatmadaw*, like any large political institution anywhere in the world, is far from omnipotent. Its dominance in politics has not managed to create broad legitimacy at home or abroad, nor has its expanded power and size created a seamless or wholly unified institution. This chapter will explore the gap between the senior officers—who regularly perform acts of *kyay-zu*[1] (which I translate in this context as 'good deeds')—and the rest of the military. Among their many often thankless tasks, soldiers and junior officers find themselves responsible for producing millions of *kyet-su*[2] (physic nuts) to generate bio-energy.

The other seams

Before proceeding, let me explain why I am not focusing on the other, better known seams in the military. For nearly 18 years now—since the current version of military governance began—there have been constant reports of imminent splits in the military. Many democracy and peace advocates have pinned hopes on internal and élite military struggles to spark some kind of dramatic and radical (by local standards) reform. Observers have focused on splits such as the hard-liners against the soft-liners; regional commanders against headquarters; junta chair, Senior General Than Shwe, versus vice-chairman, General Maung Aye, regional commanders against headquarters, and so on.

Most attention focuses on power struggles within the uppermost reaches of the regime. Until the 2004 sacking of Secretary-1 General Khin Nyunt, along with nearly all the senior officers closely associated with him, regime affairs reportedly were widely explained through a monocular lens of a single power struggle between Khin Nyunt and State Peace and Development Council (SPDC) Vice-Chair, Maung Aye. The intransigent latter was said to counter every 'baby step' by Khin Nyunt at least marginally in the direction of reform, including the dialogue with Aung San Suu Kyi and the negotiation of cease-fire arrangements with more than 20 former rebel organisations. With the sidelining of Khin Nyunt, so-called hard-liners are said to have resoundingly defeated the soft-liners. After the defeat of Khin Nyunt, the struggle is said to pit Maung Aye against his boss, junta chair, Senior General Than Shwe. At stake now seems to be less a hard-line or soft-line policy orientation than raw power, as Maung Aye has spent 13 years as Than Shwe's deputy and is thought to be chafing at the bit to ascend to the top position.

Accordingly, in a style reminiscent of Cold War Kremlinology, regime watchers scrutinise every promotion and reassignment for hints of the apparent balance tipping towards one or the other. The reshuffles of command assignments in May 2006 were widely reported in the press as a clear sign that 'Burma's Vice-Chairman [is] losing [his] grip on power'.[3] Why? Because regional commanders with close ties to Maung Aye (his *ta-bye*, or followers) were demoted. The misfortunes of the *ta-bye* are said to reflect a weakening of the power of the *hsaya* (teacher, or benefactor in this case). Their demotions involved reassignment to either ministerial or War Office appointments in Rangoon/Yangon. While it is true that getting kicked upstairs in this fashion has to represent a demotion from the quite powerful position of a regional command, not all kicks upstairs are equal or constitute an unequivocal loss of face and power for the *hsaya*.

One sure demotion and loss of influence did come in the assignment of Major-General Maung Maung Swe, Coastal Region Commander and Maung Aye's brother-in-law, as Minister of Social Welfare, Relief and Resettlement (concurrent with Minister of Immigration and Population).[4] Major-General Ye Myint, however, Eastern Commander and widely considered a Maung Aye follower, was transferred to the War Office, where he is now chief of Military Affairs Security.[5] In the past, many thought that Than Shwe placed only his most trusted generals in the War Office. It is certainly possible that he has brought in a Maung Aye follower in order to keep an eye on him, but isn't it also possible that Maung Aye had some say in this assignment? Assuming that a personnel chess game is really what is going on, perhaps Maung Aye's *ta-bye* will be able to feed him information on what Than Shwe is up to. In other words, the tea leaves of *ta-bye/hsaya* fortunes can sometimes be read in different ways.

A second set of intra-military tensions has received ample attention as well, including from myself in the past. These tensions are rooted in the enormous power and influence that regional commanders have attained in the past 18 years. In charge of all military and administrative affairs in their regions, they at times have acted like incipient war-lords, particularly in the early 1990s. While they remain very powerful, the junta and the War Office in Yangon have established formal and informal mechanisms to rein them in, starting in the early 1990s with moves to require regional commanders to serve as members of the junta and, subsequently, regular reassignments of regional commanders to War Office and Cabinet positions. Several major reshuffles have occurred since then, without producing any significant challenges to the regime. Regional commanders remain powerful, but the junta chair retains ultimate authority.

In a couple of generations, historians could find that this litany of tensions is notable not for their existence but for their irrelevance in terms of macro-level political change. Another seam, however, could have more durable structural (though not necessarily immediate political) significance in the long run: this is the growing gap between rich and poor inside the army or between senior-level officers and the sprawling and relatively impoverished rank and file and junior members of the officers' corps. To be clear, however, I do not want to overstate the consequences of this seam or any others in the *tatmadaw*. Senior military leaders have proven effective at managing conflict in the past. For example, in the case of the tensions between the junta and the regional commanders in the early 1990s, the former initiated successful reforms to patch up that seam. It is possible that the military brass could direct some portion of its pending natural resource windfalls to ameliorating the difficult conditions in which the rank and file live. Even if the leadership does not address this problem head-on, however, the common experience of poverty, or at least of diminished economic expectations, is probably unlikely—in the short term—to create sympathies and linkages between the equally impoverished civilians and soldiers, or at least not linkages significant enough to spark major political reform. The rich–poor gap in the military, however, is likely to influence Burma/Myanmar further down the road, as rent-seeking by officers will continue to hamper economic development efforts. Additionally, impoverishment among soldiers leaves them few opportunities for economic gain other than plying their skills in violence.

The gap

The *Tatmadaw* has changed quite a bit in the past decade and a half. Among the myriad changes to the institution is this vastly growing gap between the fortunes of the senior ranks of the officer corps and everyone else in the military. In 2006, some officers lived lives of unprecedented, though relative (by Burmese standards), luxury, with ample opportunities for wealth accumulation, status and *kyay-zu* performances, according to Buddhist precepts.

Let me illustrate with an example: in 1991–92, while I was doing my PhD research on military history in the military archives of Burma/ Myanmar, a quite senior colonel—one widely thought to be a *ta-bye* of the then junta chair, Senior General Saw Maung—was assigned responsibility for watching over me. Mostly, he left me alone, but once or twice he drove me from one place to another. He drove what was then widely known as a 'colonel car'. It was a Mazda 323, had no air-conditioning and the windows were stuck in a closed position, a happy condition during the rainy season, but utterly suffocating during the hot season. Nonetheless, it was a symbol of great status in a poor country with few cars, little opportunity for wealth accumulation and relatively low expectations. At the same time in other Southeast Asian countries—such as Thailand and Indonesia, for example—colonels on active duty drove nice Toyota sedans or other up-market cars. In Burma/Myanmar, however, wealth and status were measured on a far more limited scale.

Today, colonels, like rock stars and business owners in Burma/ Myanmar, wouldn't be caught dead in a Mazda 323.[6] Now they drive at least what their counterparts in other Southeast Asian countries drive.[7] In 2006, the senior officer corps of the *Tatmadaw* lived far different lives from that of their predecessors, even those serving under the same military regime only a decade or so earlier. As weak as the Burmese economy is, it has produced opportunities for the scaling up of wealth/status measures across the board—in pop culture, business and the military.

The life of a senior officer is now typically one of great comfort, possibility and *kyay-zu*. Today's older, active-duty senior officers paid their dues on the battlefields of Burma's two generations of civil wars, and the mid-career officers were promoted for their service in the difficult reconstruction of the State in the post-1988 period. The past 18 years have provided extensive opportunities for accumulating wealth—not always for themselves, but certainly for their families and entourages. Along with those opportunities for wealth accumulation came chances to ascend (in some cases quite quickly) the ladder of social status. Like non-military *lu-gyi* ('important people' or 'big shots') in Burma/Myanmar, large numbers of officers have now become quite visible public figures in ways they weren't before 1988. They regularly show off their high status, driving around in expensive cars, eating at expensive restaurants, promoting their children in business or educational sectors, practising *kyay-zu* by providing largesse to monks and pagodas, inscribing their names on donation plaques at religious tourist sites and—most importantly—reminding the population and rank and file just who is in charge.

For the rank and file, however, the situation is comparatively grim, though a different kind of grimness from that existing 10 or 15 years ago. Soldiers live in some ways as soldiers always have, with livelihoods formally above, but not always far above, the poverty line. Opportunities for advancement and livelihood enhancement continue to depend on the goodwill of their commanding officers. Their lives remain difficult, although unlike before 1988, their lives are less frequently on the line during their service. Instead of constant fighting, they are tasked with the thankless job of providing the muscle to build a repressive state. Before 1988, soldiers cycled through frequent combat assignments, usually far away from their homes and often in areas where they didn't speak the same language as the locals or had little logistical support to sustain them. Given the outdated and weak equipment of the pre 1988 *Tatmadaw*, tens of thousands of them died in battles with a plethora of different armed groups fighting against the State. From 1989, many of these armed groups collapsed from within, concluded cease-fire agreements with the government or surrendered. As combat contracted, the military inexplicably expanded.

Post 1988, soldiers thus found themselves in a new terrain, metaphorically (*kyet-su* plantations) and geographically. Most are no longer required to fight insurgents on a regular basis, and some have found themselves serving in units closer to home or at least are stationed at distant garrisons where they are not likely to be shot at. Their jobs involve new kinds of assignments, ranging from corralling local people into service on infrastructure and construction projects and collecting money at road check-points, to (in the past two years) planting or pressing others into planting the infamous *kyet-su* trees all over Burma. Although the *Tatmadaw*'s acquisition of higher-tech weaponry since 1988 has ushered in a revolution (by Burmese standards) in military affairs, its institutional development has frequently failed to keep pace with the demands of sustaining its vastly larger rank and file. In other words, no comparable revolution in military social affairs has taken place.

Kyay-zu: explaining the rising fortunes of senior officers

Two major changes in the political and economic environment of Burma/Myanmar account for the growth of opportunities for wealth accumulation for very senior officers. One is the explosion of rent-seeking opportunities that has emerged with the state rebuilding process from the late 1980s until today. The second was that changes in the world economy ushered in at the end of the Cold War—and particularly the expansion of neo-liberal institutions and policies—gave senior officers access to an unprecedented range of lucrative (and often informal) business partnerships.

First, the military take-over in 1988 was the take-over of only the shell that remained of the socialist state. The State Law and Order Restoration Council (SLORC) began a sprawling range of nation and state-building activities designed to maximise order and modernise Burma/Myanmar. To carry out this massive and often uncoordinated set of state rebuilding programs, the junta undertook a huge expansion of the armed forces. From 1988 to 1996, the *Tatmadaw* probably doubled in size.[8] Local commanders in towns and villages throughout the country seized land to construct new army garrisons, while the numbers of naval and airforce bases also increased (Selth 2002). The military also expanded its economic and industrial base, and set up lucrative military corporate ventures, such as agricultural plantations, banks and holding companies such as the Union of Myanmar Economic Holdings Limited (UMEHL).

The junta delegated the day-to-day administration of this emerging behemoth to its regional commanders. Regional commanders have supervised the construction of roads, housing, suburbs and markets; rearranged and displaced urban and rural populations to accommodate tourism, military expansion and other state priorities; and expanded surveillance and crowd-control capabilities. With administrative, military and political jurisdiction over their geographically vast command areas, regional commanders have amassed enormous wealth and power, especially when posted to the commands flanking Burma/Myanmar's borders with China and Thailand. There, they oversee formal and informal trade, investment, transport and border crossings—all of which provide ample opportunities for personal and institutional enrichment.

The expanded presence of the *Tatmadaw* is visible throughout the country. In addition to the ever-present but by now often peeling red-and-white propaganda billboards exhorting locals to support the Rangoon-based state, evidence of increased military presence includes (but is not limited to)

- massive expansion of army garrisons, often set up on land requisitioned from farmers or local businesses with little or no compensation
- the initiation of large, flashy infrastructure construction projects—such as oil pipelines, microwave stations, universities and hydroelectric dams—that typically rely on conscripted local labour and taxation[9]

- increased numbers of road check-points or toll gates at which proceeds benefit army units, as well as local Union Solidarity Development Association (USDA) groups or line ministries

- increased kinds and amounts of business licence fees and levies on all civilians. Most are payable either directly to military units or indirectly channel some of the money to the *Tatmadaw* through USDA, police and line ministry offices. According to the Karenni Development Research Group (2006:41), since 1988, Kayah State has been subject to 'porter fees, gate fees, military fund contributions, sports fees, road and bridges fees, fire sentry fees, labour contribution fees, and levies on farms, farm water, and crops'

- direct army ownership of plantations and agricultural land (usually marked by official signage), where nearby villagers are expected to 'donate' their labour

- increased pressures on farmers to expand areas of cultivation or plant crops defined by the SPDC as national priorities

- expansion of the number of model villages. In some areas, such as northern Rakhine, these approximate strategic hamlets— arrangements that are largely involuntary for social control of potentially hostile populations (usually ethnically defined) or involve resettlement of displaced populations for counter-insurgency purposes. In other areas, villagers apply for model village status to obtain some government services in exchange for adhering to strict planting and production schedules set by local and ministerial officials from the Ministry for Progress of Border Areas or the relevant line ministries.[10]

Commanding officers, who oversee all of these major state-building enterprises, thus have access to a wide range of daily rent-seeking opportunities of which they can take advantage—either directly for themselves or their directorate or command or indirectly by their families or *ta-bye* for a variety of different purposes. It appears, however, that the greatest sums have flowed to the top of the chain of command largely because of changes in the global environment.

The world economy

To an unprecedented degree, populations living in Burma/Myanmar have been affected by the increased ease with which capital, legal and illegal commodities and people could move in the late twentieth and early twenty-first centuries. Since the British era, the formal financial system of Burma/Myanmar has never reached most Burmese in the central or border regions. Most instead rely on friends or families for credit or on money-lenders charging extremely high interest rates. According to Turnell (2006), 80 per cent of the country's farmers have no access to banks or other forms of formal credit whatsoever. Moreover, given the decades-long civil wars that have occurred in the border regions, the country has always been home to shadow economies and trans-border networks of brokers, traders, money-lenders, traffickers and militias.

In the past decade, these regions have seen a dramatic deepening and thickening of these networks. What is different now is that the deregulation of much of the world's financial system since the 1980s has broken down many barriers to illegal and legal trade in the commodities produced in the country's resource-rich border regions. Additionally, the cosying up to China of former General Khin Nyunt and his decision to grant National Registration Cards to Kokang Chinese within Burma/ Myanmar has probably hastened the pace of formal and informal Chinese investment in and exploitation of natural resources. In the states sharing a border with Thailand, there has been a considerable increase since 1988 of traffic going out, as young people especially traverse the porous border to seek work either in the many sweatshops and factories along the border or in Chiang Mai and Bangkok.[11] Globalisation has thus brought about a considerable transformation in social relations in the border states as well as the rest of the country.

While senior officers do not directly run or control these networks of economic linkages, they have certainly benefitted from the reinsertion of Burma/Myanmar into the world economy. Wives, sons, daughters, in-laws and cousins of senior officers have seized business opportunities that involve the purchase of undervalued land, gem production, hotels and tourism businesses, monopolistic access to economic assets and extortionary joint-venture requirements for foreign investors.

Whereas the families and *ta-bye* of *lu-gyi* expected comfortable but not necessarily luxurious living standards in the past, now they expect nothing less than the mass consumerist luxuries that they imagine élites everywhere else have. They also expect and demand new degrees of deference. They often accompany senior officers to the frequent processions through towns and villages where the officers open roads, schools, plantations or bridges or make donations at pagodas (often with cash appropriated informally from businesses). Local government officials and army commanders oversee the production of these pageants and assure that villagers and townspeople exaggerate their gratitude and loyalty to the *lu-gyi*. Few inside the country can mistake the *kyay-zu* performed by senior officers in the past 18 years.

Kyet-su: the misfortunes of soldiers

Mismanagement of the expansion of the military and the economy led the War Office to declare in the late 1990s a policy of self-reliance for local military units. At that moment, military leaders threw up their hands at the logistical nightmare they had created and directed local units to raise operating revenues from whatever the local economy could provide. Self-reliance, however, did not mean much in the way of autonomy, as local units remained subject to macroeconomic, fiscal and planning policies—like the required cultivation of *kyet-su*—that often redirected soldiers' efforts away from whatever economic efficiencies they might have been able to achieve.

The *Tatmadaw* grew from 180,000 in 1988 to about 300,000 in the mid 1990s. There was and is no official military conscription in Burma/Myanmar. In the early years of this junta, some families sent their sons off to join the military, often attracted by promises of access to subsidised petrol, rice and cooking oil. Other, very poor families sent sons to join the military because they had no alternative. Additionally, various levels of military officials have periodically set quotas for township and village leaders to send young men to the rapidly expanding military. Nonetheless, there is ample anecdotal evidence that today many units (including those involved in combat) are considerably under-strength and that unit commanders have great difficulty recruiting and preventing desertion.[12]

The expanded range of economic opportunities that accompanied the so-called opening of the economy after 1988 has brought resources into the *Tatmadaw*, but relatively few are directed at improving the lives of the soldiers. Instead, military leaders have prioritised the purchase of an extensive (though not necessarily integrated) range of modern weapons from abroad (Selth 2002). Moreover, they have created an almost parallel state service sector that provides relatively high-quality opportunities for health care, education and other social services for the officer corps and their families. In some parts of Burma/Myanmar, soldiers' rations, allowances and wages are not enough for a single soldier to live on, much less to support their families. As a result, underpaid soldiers can feed their own families only by participating in the informal and illegal economy, levying informal taxes, stealing villagers' harvests and collecting unauthorised road tolls.

In mid 2006, the SPDC announced a raise in salaries for all government servants including those in the military. It subsequently undertook a crack-down on informal wealth generation by civil servants, starting in the customs department. It is not clear whether the salary increases are adequate for the needs of most soldiers, or whether indeed they will ever receive the full raises. Unit commanders themselves are reported to tax soldiers' wages to finance unit requirements (such as the purchase of shares in UMEHL, often required by regional commanders) as well as their own personal needs. Additionally, in the 2006 salary hikes, soldiers' raises were proportionately smaller than those of more senior officers.

Why hasn't the War Office done more to address the impoverishment of its rank and file? As seems to happen in other agencies of the SPDC state, junior and mid-career officers probably are disinclined to report problems up the chain of command. If the policy of the junta and the War Office is one of local self-reliance, that represents an order to be carried out, not one to be renegotiated because it is difficult to carry out. Battalion commanders might have opportunities to raise their concerns about this at quarterly regional command meetings, but it seems unlikely that regional commanders would transmit this information up the military chain of command given that it reflects poorly on their management of troops and their areas of operation.

Implications of the gap

To be clear, I am not suggesting that the widening of the gap between the very rich in the *Tatmadaw* and the rank and file will finally be the internal split that will bring down the military government. Throughout its several decades of rule, the *Tatmadaw* has weathered many internal problems and personal infighting, but the leadership has always successfully held it together. Junior officers might bristle at the current inequities; however, for 18 years under SLORC/SPDC management, these same junior and mid-career officers have regularly been promoted into positions where they can take their shot at the remaining spoils, rather than upending the system that served them poorly at lower ranks.

At some point, however, future governments of any stripe—military or civilian, authoritarian or democratic—will inevitably need to confront the problems of a system of wealth generation that directs state resources into unproductive endeavours and enterprises. Any move to a political system that enshrines security of contract and rule of law will inevitably weaken the patron–client, entourage modes of dividing up the thus far expanding spoils (likely to continue to expand in the near future, at least in the natural resources and energy sectors). The maintenance of such a large, unwieldy force structure of about 300,000 will also be a drain on public coffers, and it seems inevitable that at some point, the *Tatmadaw*, like the rest of the armed groups[13] operating inside the country's borders, will have to demobilise many soldiers. Such a move will unleash tens of thousands of men trained in violence into an economy incapable of absorbing these newly jobless.

Notes

1 This can be translated as 'good deed' or 'benefit', or sometimes 'gratitude'.
2 Physic nuts or castor-oil trees. The Burma/Myanmar government has been promoting the cultivation of physic-nut trees throughout most of the nation. According to Brigadier-General Hla Htay Win, Yangon Regional Commander and Chair of the Yangon Division PDC, 'Physic nut oil can be used to meet the fuel needs of the nation to some extent and it will be useful for the people in the long run and it is necessary to grow the plant widely throughout our country' (*New Light of Myanmar*, 8 February 2006). The newspaper reported that Yangon Division alone was planning to collect grafts and cultivate 500,000 acres of *kyet-su* between 2006 and 2009.
3 Headline from Democratic Voice of Burma radio, 16 May 2006.
4 It should be noted that immigration is an increasingly powerful portfolio, given the expanded numbers of Burmese workers now emigrating (or being exported) for labour abroad.
5 Military Affairs Security has assumed many of the responsibilities of Khin Nyunt's military intelligence departments (which were disbanded at the time of his sacking).
6 A stark example of the elevation of expectations can be seen in a *Myanmar Times* story (Puii 2006:29), which reports that musician DJ Thxa Soe 'says he drives a pick-up instead of a Land Cruiser because all the profits from his music are eaten up by VCD piracy'.
7 Officially, the military allows colonels to drive saloon cars (sedans) with the military-star licence or number plate. Privately, they can purchase and register their own cars. Most colonels are probably not driving makes such as Mercedes and Lexus, but they and their families are seen around Yangon and other areas in four-wheel drives and other up-market automobiles.
8 In its first decade, the junta also spent more than $1 billion on 140 new combat aircraft, 30 naval vessels, 170 tanks, 250 armoured personnel carriers, as well as rocket-launch systems, anti-aircraft artillery, infantry weapons, telecommunications surveillance equipment and other hardware. See Brooke 1998 and Davis and Hawke 1998. See Tan 2004 and Aung Zaw 2006 for updates.
9 See, for example, EarthRights International 2001 and Karenni Development Research Group 2006.
10 Human Rights Documentation Unit 2003; Loo 2004:168–9; Amnesty International 2004:22–4.
11 From General Chaovalit's visit with Senior General Saw Maung after the September 1988 coup through to the mid 1980s, Thai companies landed the

lion's share of timber concessions along the border between the two countries. Since then, Thai investment has waxed and waned, given competition with other sources of capital (China, India) and fluctuating relationships between the junta and successive Thai governments.

12 See, for example, ALTSEAN 2005; Amnesty International 1989a, 1989b, 1991, 1996; Global Witness 2003, 2005; Karen Human Rights Report (various issues of News Bulletin at http://www.khrg.org); Project Maje 1996; Risser et al. 2004; Sakhong 2003; *Shan Herald Agency for News* 2005; Smith 1991; Thailand Burma Border Consortium 2005.

13 Here, I am thinking of the ethnic armed cease-fire groups, who have been allowed to retain their weapons and sustain armies.

References

ALTSEAN, 2004. *Burma Briefing: issues and concerns*, Alternative ASEAN Network on Burma, Bangkok.
——, 2005. *Interim Report Card: a summary of political and human rights developments in Burma, July 2004–February 2005*, Alternative ASEAN Network on Burma, Bangkok.
Amnesty International, 1989a. *Myanmar (Burma): call for dissemination and enforcement on the use of force*, Amnesty International, London.
——, 1989b. *'No Law at All': human rights violations under military rule*, Amnesty International, London.
——, 1991. *Myanmar (Burma): continuing killings and ill-treatment of minority peoples*, Amnesty International, London.
——, 1996. *Myanmar: human rights violations against ethnic minorities*, Amnesty International, London.
——, 2004. *The Rohingya Minority: fundamental rights denied*, Amnesty International, London.
Aung Myo, M., 1999. *Military doctrine and strategy in Myanmar: a historical perspective*, SDSC Working Papers, Canberra.
Aung Zaw, 2006. 'A growing tatmadaw', *The Irrawaddy*, Chiang Mai.
Bamforth, V. Lanjouw, S. and Mortimer, G., 2000. *Conflict and Displacement in Karenni: the need for considered responses*, Burma Ethnic Research Group, Chiang Mai.
Beyrer, C. Mullany, L. Richards, A. Samuels, A. Suwanvanichkij, V. Lee, T. and Franck, N., 2006. 'Responding to AIDS, TB, malaria and emerging infectious diseases in Burma: dilemmas of policy and practice', report

for the Center for Public Health and Human Rights, Johns Hopkins Bloomberg School of Public Health, Baltimore. Available from: http://www.jhsph.edu/humanrights/burma_report.pdf.

Brooke, M., 1998. 'The armed forces of Myanmar', *Asian Defence Journal*, January:13.

Davis, A. and Hawke, B., 1998. 'Burma: the country that won't kick the habit', *Jane's Intelligence Review*, 10 (March):26–31.

EarthRights International, 2001. *Fatally Flawed: the Tasang Dam on the Salween River*, EarthRights International, Chiang Mai.

Fink, C., 2000. 'An overview of Burma's ethnic politics', *Cultural Survival Quarterly*, 24(3). Available from http://www.cs.org/publications/CSQ/csq-article.cfm?id=1015

——, 2001. *Living Silence: Burma under military rule*, White Lotus, Bangkok.

Global Witness, 2003. *A Conflict of Interests: the uncertain future of Burma's forests*, Global Witness, London.

——, 2005. *A Choice for China: ending the destruction of Burma's northern frontier forests*, Global Witness, London.

Havel, V. and Tutu, D., 2005. *Threat to Peace: a call for the UN Security Council to act in Burma*, DLA Piper Rudnick Gray Cary, New York.

Heidel, B., 2006. *The Growth of Civil Society in Myanmar*, Books for Change, Bangalore.

Human Rights Documentation Unit, 2003. *Burma Human Rights Yearbook*, National Coalition Government of the Union of Burma, Nonthaburi, Thailand.

Human Rights Watch, 2006. 'Human Rights Watch World Report 2006', Human Rights Watch, New York.

International Crisis Group, 2004. 'Myanmar: aid to the border areas', *Asia Report*, No. 82, International Crisis Group, Yangon/Brussels. Available from http://www.crisisgroup.org/library/documents/asia/burma_myanmar/082_myanmar_aid_to_the_border_areas.pdf.

Kachin Independence Organisation, 1995. 'Kachin Resettlement Report', Kachin Independence Organisation.

Karen Human Rights Group, 2006. 'Report from the Field: abuses in SPDC-controlled areas of Papun District', report by Karen Human Rights Group. Available from http://www.khrg.org/khrg2006/khrg06f3.pdf .

——. *Karen Human Rights Report*, various issues of News Bulletin, Karen Human Rights Group. Available from http://www.khrg.org.

Karenni Development Research Group, 2006. *Dammed by Burma's Generals: the Karenni experience with hydropower development from Lawpita to the Salween*, Karenni Development Research Group.

Lahu National Development Organisation, 2002. *Unsettling Moves: the Wa resettlement program in eastern Shan State, 1999–2001*, Lahu National Development Organisation, Chiang Mai.

Lintner, B., 1994. *Burma in Revolt: opium and insurgency since 1948*, Westview Press, Boulder.

Loo, N.J., 2004. 'Myanmar', in T. Onchan (ed.), *Non-Farm Employment Opportunities in Rural Areas in Asia*, Asian Productivity Organisation, Tokyo:164–78.

Mon Forum, 2005. *Past and present suffering of civilians in Yebyu Township under the name of security to gas pipelines*. Available from http://www.rehmonnya.org/report_detail.php?ID=17 (accessed 11 June 2006).

Network for Democracy and Development, 2006. *The White Shirts: how the USDA will become the new face of Burma's dictatorship*, Network for Democracy and Development, Mae Sariang.

Puii, Z., 2006. 'No Land Cruiser for DJ Thxa Soe', *Myanmar Times*, 5–11 June 2006:29.

Project Maje, 1996. 'Dacoits, Inc.' Report by Project Maje, Portland. Available from: http://www.projectmaje.org/pdf/dacoits.pdf.

Human Rights of Monland (Burma) 2005. 'USDA's forced registration and preparation for future election', 30 October. Available from http://www.rehmonnya.org.previousnewsdetails.php?ID=38 (accessed 11 June 2006).

Risser, G., Kher, O. and Htun, S., 2004. 'Running the Gauntlet: the impact of internal displacement in southern Shan State', Humanitarian Affairs Research Project, Asian Research Centre for Migration Institute of Asian Studies, Chulalongkorn University, Bangkok. Available from: http://www.ibiblio.org/obl/docs3/Gauntlet-ocr.pdf.

Sakhong, L.H., 2003. 'Human rights and the denial of minority rights in Burma', paper presented at the Sixth International Conference on Human Rights, Manila.

Selth, A., 2002. *Burma's Armed Forces: power without glory*, EastBridge, Norwalk, Connecticut.

Shan Herald Agency for News, 2005. 'Show business: Rangoon's "war on drugs" in Shan State', *Shan Herald Agency for News*. Available from http://206.225.87.155/live/shan/

——, 2006. 'Ceasefire factions keep up "uncivil" war', *Shan Herald Agency for News*. Available from http://206.225.87.155/live/shan/

Shukla, K., 2006. *Ending the Waiting Game: strategies for responding to internally displaced people in Burma*, Refugees International, Washington, DC.

Smith, M., 1991. *Burma: insurgency and the politics of ethnicity*, Zed Books, London.

——, 1994. *Ethnic Groups in Burma: development, democracy and human rights*, Anti-Slavery International, London.

——, 2002. *Burma (Myanmar): the time for change*, Minority Rights Group. Available from http://www.minorityrights.org/

South, A., 2003. *Mon Nationalism and Civil War in Burma: the golden sheldrake*, RoutledgeCurzon, London.

Steinberg, D.I., 1997. 'The Union Solidarity & Development Association: mobilization and orthodoxy', *Burma Debate*.

——, 2001. *Burma: the state of Myanmar*, Georgetown University Press, Washington, DC.

Tan, A., 2004. *Force modernization trends in Southeast Asia*, Working Paper, Institute for Defence and Strategic Studies, Singapore.

Taylor, R., 1987. *The State in Burma*, University of Hawai'i Press, Honolulu.

Thailand Burma Border Consortium, 2005. *Internal Displacement and Protection in Eastern Burma*, Thailand Burma Border Consortium, Bangkok.

Turnell, S., 2006a. 'Burma's economic prospects', Senate Foreign Relations Subcommittee on East Asian and Pacific Affairs, Washington, DC.

——, 2006b. 'Burma's economy 2004: crisis masking stagnation', in T. Wilson (ed.), *Myanmar's Long Road to National Reconciliation*, Institute of Southeast Asian Studies, Singapore.

Unrepresented Nations and Peoples' Organization, 2005. *The military is plundering Burma's forests*, Unrepresented Nations and Peoples' Organization. Available from http://www.unpo.org/news_detail.php?arg=39&par=3281 (accessed 11 June 2006).

Wain, B., 1999. 'Myanmar seeks stability in the hinterland', *The Asian Wall Street Journal*, 1 May.

4 Conflict and displacement in Burma/Myanmar

Ashley South

Patterns of forced migration in Burma/Myanmar are structured by the changing nature of conflict in the country. While acutely vulnerable internally displaced persons do live in those few areas of the country that are still affected by significant armed conflict (especially in the insurgent-prone eastern borderlands), the phenomenon of forced migration is more widespread and complex. Yet assessments of forced migration in the country as a whole have tended to be obscured by the focus on parts of eastern Burma that are accessible to agencies working across the border from Thailand. Much less is known about the situation in other geographic areas, or about displaced populations not accessible to the armed opposition groups with which cross-border aid agencies cooperate. Another problem is that the literature on the political economy of conflict and displacement is sparse, and the majority of investigators have been constrained by their own sociopolitical agendas. Their emphasis on 'problem-finding' has not taken account of the positive trends that have emerged in the past decade.

This chapter attempts to redress the balance of existing research by addressing forced migration in parts of the country that are not readily accessible from the Thai-Burma border. It identifies new forms of forced migration that have emerged with the existence of cease-fires in many previously armed conflict-affected areas—which could be expected to occur in other affected areas if or when insurgency ends along the border. A better understanding of the situation in areas that are no longer affected by armed conflict could help to prepare local and international actors for future developments in areas that are currently beset by the State's counter-insurgency operations. In many situations, migration itself constitutes a coping mechanism—as illustrated by the variety of rezones labelled 'economic migration'. Towards this end, the study incorporates rights-based perspectives, but also adopts an actor-oriented perspective, focusing on the agency of displaced people rather than viewing them as passive victims. It seeks to identify the positive responses of individuals and communities to the problems they face.

Terminology and typology

In this chapter, forced migration is conceptualised as a subset of population movement in general, and internal displacement is a division of forced migration. The *Guiding Principles on Internal Displacement* (UNHCR 1998) define internally displaced persons as 'persons or groups of persons who have been forced or obliged to flee or to leave their homes or places of habitual residence, in particular as a result of or in order to avoid the effects of armed conflict, situations of generalised violence, violations of human rights or natural or human-made disasters, and who have not crossed an internationally recognised State border' (UNHCR 1998).

This chapter identifies and describes three main types of forced migration in and from the country, each of which is presented with reference to material drawn from different geographic areas (Table 4.1).

The first type is armed conflict-induced displacement, which occurs either as a direct consequence of fighting and counter-insurgency operations or because armed conflict has directly undermined human

Table 4.1 Typology of forced migration

Internally displaced persons	Other forced migrants	
Type 1	Type 2	Type 3
Armed conflict-induced	State-society conflict-induced (post-armed conflict)	Livelihoods vulnerability-induced ('distress migration')

and food security. Type One forced migration is linked to severe human rights abuses across Karen State, in eastern Tenasserim Division, southern Mon State, southern and eastern Karenni State, southern Shan State and parts of Chin State and Sagaing Division.

The second type is state–society conflict-induced displacement, which is generally post-armed conflict and caused by military occupation and or 'development' activities. Type Two forced migration could be due, for example, to land confiscation by the *Tatmadaw* (military) or other armed groups, or it could be caused by infrastructure construction. It could also be a product of predatory taxation, forced labour and other abuses. All of the border states and divisions are affected by militarisation and/or development-induced displacement, including Arakan (Rakhine) and Kachin States, as well as many urban areas. Type One and Type Two forced migrants are internally displaced persons whose displacement is the result of conflict—either active, armed (Type One) or latent conflict, or the threat of the use of force (Type Two).

A third type is livelihood vulnerability-induced displacement, which is the primary form of internal and external migration in and from Burma. Main causes include inappropriate government practices and policies, limited availability of productive land, poor access to markets resulting in food insecurity, lack of education and health services and stresses associated with transition to a cash economy. Type Three displacement occurs across the country, especially in remote townships. Type Three movements involve a particularly vulnerable subgroup of economic migrants and result from limited choices faced by marginal populations. As such, they constitute a form of forced migration. Migration due to opium-eradication policies is included under Type Three because the causes of the movement are related to livelihood issues; with the exception of some Wa areas, people are not ordered to move (opium eradication-induced migration could, however, also be considered under Type Two forced migration, due to the forcible nature of the opium bans, the severe shock to livelihoods involved and the links to development activities).

There are important links between these three types of displacement, each of which undermines traditional livelihood options and depletes people's resource base. Type One characterises zones of continuing armed conflict and some adjacent areas; Type Two is particularly prevalent in remote and underdeveloped conflict-affected areas where cease-fires have been agreed, and also affects urban relocatees; Type Three is characteristic of remote areas, particularly those where armed conflict has ceased. This progression in causes of population movement is not strictly linear: many people are in cyclical transition between different phases of displacement and could be categorised in different ways at different times.

Internally displaced persons: population estimates

For many Burmese citizens, patterns of often cyclical migration involve periods spent as labourers in other countries and/or more extended periods as refugees in neighbouring countries. The causes and other aspects of population movements within Burma (internal migration) and beyond its borders (external migration) are closely linked and often relate to serious and systematic abuses of a range of basic rights. This chapter focuses primarily on the situation of forced migrants inside Burma.

It is difficult to assess the numbers of internally displaced persons and the scale of the problem. Counting only people who have been forcibly displaced since 2004, the number of internally displaced persons in eastern Burma will be no more than 100,000 (including 25,000 people displaced by the *Tatmadaw* in northern Karen State, since February 2006). The number of previously displaced persons for whom no durable solution has been found must, however, be calculated in the millions. Since 1996, more than 2,800 villages are known to have been destroyed and/or relocated *en masse*, or otherwise abandoned, due to *tatmadaw* activity—including at least 306 villages between 2002 and 2005 alone (TBBC 2005b). While unknown numbers of these villages have since been resettled, most remain depopulated. According to the Thailand Burma Border Consortium (TBBC) and its local partner groups, there were 540,000 internally displaced persons in eastern Burma in mid to late 2005. These figures do not include Type One internally displaced persons who choose not to make themselves available to armed opposition groups, or large numbers of people who have achieved at least semi-durable solutions to their plight. Nor do they include the hundreds of thousands of Type Two and Three internally displaced persons in other parts of Burma.

Long-term patterns of displacement have tended to be under-researched, but warrant attention because they are crucial to understanding the dynamics of conflict and patterns, impacts of and responses to forced migration in Burma. Armed conflict-induced (Type One) displacement often occurs among communities that periodically shift their location for sociocultural reasons and/or to access agricultural land. The scale of displacement in Karen and other areas in the past 50 years has, however, been out of all proportion to any traditional patterns of migration. Furthermore, forced migration among significant segments of the Karen and other ethnic nationality communities is not a one-off phenomenon. Rarely do individuals, families or communities return in a simple manner to their original location, which could have come to be occupied by the *Tatmadaw* or other hostile groups, resettled by other displaced people and/or planted with land-mines. In-depth interviews conducted in 2003–04 with a group of 36 internally displaced Karen in the Papun Hills in northeastern Karen State revealed that many had undergone more than 1,000 migration episodes. Five had been forcibly displaced more than 100 times, some dating back to the 1940s. The majority of migration episodes followed directly from fighting, because of severe human rights abuse or because armed conflict had directly undermined sustainable forms of agriculture. I consider the situation of the Karen in the following case study.

Type One forced migration: the Karen

For more than half a century, life across much of rural Burma has been profoundly affected by armed conflict. In many ethnic minority-populated areas, repeated incidents of forced displacement—interspersed with occasional periods of relative stability—have been a fact of life for generations. Those cases in which human displacement occurs as a direct result of armed conflict can be classified as Type One forced migration. The situation of the Karen provides an illustration of armed conflict-induced displacement.

The Karen community consists of a diverse collection of ethno-linguistic groups, which nevertheless share a number of common characteristics. At least two-thirds of the five to seven million Karen in Burma are Buddhists. Many of the conceptions of ethnic identity in contemporary Burma remain rooted in the pre-colonial past and in the often traumatic colonial experience (Thant 2001). The Karen ethno-nationalist movement emerged during the British colonial period, when Christian Karen élites first began to express the idea of a Karen nation, including all elements of the diverse socio-linguistic community. The Karen National Union (KNU), which went underground in January 1949, was from the outset led by educated Christian élites—in the name of all Karen. In successive years, the rebellion continued as a response to the repressive policies of successive governments in Yangon, and the perceived 'Burmanisation' of the State (Smith 1999).

In the decade after 1962, when General Ne Win's *Tatmadaw* took control of the country, the KNU and other ethnic insurgent groups received new injections of recruits from government-controlled Burma. Ne Win's disastrous 'Burmese way to socialism' also provided the insurgents with new sources of funds, as the economy collapsed and became dependant on smuggled goods—most of which came from neighbouring Thailand. The KNU and other armed ethnic groups taxed the black-market trade, allowing several rebel leaders to prosper and build up well-equipped armies. Meanwhile, the KNU and other insurgent 'liberated zones' took on some of the characteristics of *de facto* states, with military and parallel civilian administrations, and health and education systems.

This period saw the emergence of significant economic agendas in the prosecution of armed conflict in Burma. These are epitomised by the rise of the KNU's General Saw Bo Mya, a tough field commander, staunch Christian and anti-communist, who became a key asset in Thai and US strategy in the region. Like most ethnic insurgent groups, the KNU has claimed to be fighting for democracy in Burma—especially since the 1988 democracy uprising. This position has been reflected in a series of alliances struck with pan-Burma opposition groups which fled to the border areas after the events of 1988 and 1990. The democratic ideal has not, however, always been honoured in practice, and the liberated zones have often been characterised by a top-down tributary political system, aspects of which recall pre-colonial forms of sociopolitical organisation. While General Bo Mya et al. have certainly been inspired in their conflict with the central government by genuine and strongly held grievances, many insurgent commanders and their families have also benefitted financially from protracted armed conflict in Burma—especially from the taxation of black-market trade, and from natural resource extraction (in the case of the KNU, logging and mining activities).

Under General Bo Mya, S'ghaw-speaking élites from the lowlands began to unify—and dominate—Karen society in the eastern hills. This internal colonisation had unforeseen consequences, as an underclass of mostly Buddhist subalterns came to resent the domination of an increasingly corrupt and authoritarian alien élite. The end result was rebellion within the Karen nationalist ranks and the formation of the Democratic Kayin Buddhist Army (DKBA) in late 1994 (Smith 1999).

During the early 1980s, government forces gained the upper hand in the civil war and the first semi-permanent Karen refugee camps were established in Thailand, as civilians (and rebel soldiers) fled *tatmadaw* offensives along the border. By 1994, with the fall of its headquarters at Mannerplaw, the KNU was in serious trouble. The crisis was compounded by the loss of most of the remaining Karen liberated zones (in southern Karen State and Tennasserim Division) during a major dry-season *tatmadaw* offensive in 1997.

The KNU today is a greatly weakened force and no longer represents a significant military threat to the State Peace and Development Council (SPDC). The Karen National Liberation Army (KNLA) still has some 5–7,000 soldiers, deployed in seven brigades (including mobile battalions and village militias), and more than 1,000 active political cadres (including youth and women's wings). At any one time, however, about half of these personnel are located among the 148,000 refugees living in 10 camps (seven Karen, two Karenni, one Shan) in Thailand.

Although the KNU is in danger of becoming marginalised on the Burmese political stage and as an arbiter of Karen affairs, its continuing symbolic importance cannot be denied. The KNU is the oldest and, to many Karen people and Burma watchers, the only legitimate Karen ethno-nationalist group. Having fought for independence (and later, autonomy) from Yangon since 1949, and not having followed other armed ethnic groups into the cease-fire movement, the KNU retains strong credibility in opposition circles.

After more than half a century, armed conflict in Burma has thus become institutionalised and associated with deep-rooted political economies. Commanders on both sides of the front lines (including those, such as the DKBA, which have agreed cease-fires with the government) often rely on the taxation of black-market goods, extraction of natural resources (logging and mining) and other unregulated practices (including the drug trade) to enrich themselves and their retinues, and to support the armed groups, control of which brings the power to extract further 'tribute' and political power—a vicious circle.

The prevalence of such greed-based models of conflict world-wide tends to provoke scepticism of élite claims to represent ethnic communities. This is especially the case among international agencies and observers with experience of armed conflict and its impacts in other parts of the world, who tend to focus on greed models and the political economy of conflict in Burma. Such perspectives, however, under-appreciate the (often contested) legitimacy of many insurgent and cease-fire groups and underestimate the levels of support they enjoy in their constituencies. In contrast, opposition supporters (especially those based outside Burma) tend to emphasise the struggle against a repressive regime and 'justice/legitimate grievance' models of conflict, and are often supportive of élite-generated ethno-nationalist agendas, without questioning whose interests they serve.

Burma's ethnic insurgent groups have positioned themselves as the defenders of minority populations against the aggression of state forces. They have adopted guerrilla-style tactics, which have invited retaliation against the civilian population, but against which the armed groups have been unable to defend villagers. Since the 1960s, in response to protracted insurgencies in most ethnic nationality-populated areas, state forces have pursued often brutal counter-insurgency strategies, including the forced relocation of civilian populations deemed sympathetic to armed ethnic and communist groups (Taylor 1985). The KNU and other insurgent groups have an interest in controlling, or at least maintaining, civilian populations in traditional Karen lands—as a source of legitimacy, and of food, intelligence and soldiers, porters and so on. Therefore, KNU cadres regularly organise village evacuations to 'protect' villagers from *tatmadaw* incursions (a service that is appreciated by many internally displaced persons). Clearly, the KNU and other insurgent organisations bear some responsibility for the plight of civilians in areas where they operate. For nearly 60 years, they have pursued an armed conflict against the central government, although the possibility of any military victory probably disappeared during the 1970s—or, at the latest, after the fall of the last KNU liberated zones in the mid 1990s.

Such complexities notwithstanding, most forms of forced displacement—and associated serious human rights abuses—still occur in the context of the *Tatmadaw*'s 'four cuts' counter-insurgency strategy (and, more recently, as a result of the activities of government-aligned militias). Having issued orders to relocate to areas firmly under state control, *tatmadaw* columns often return to remote areas that have been 'cleared' to ensure that they are not resettled (which they often are): many villages are therefore 'serially displaced'.

It is therefore not surprising that armed conflict and counter-insurgency operations in rural Burma have severely disrupted traditional ways of life. Most of the rural and peri-urban population of eastern Burma has been displaced or otherwise affected at some point during the past 50 years. Since the late 1980s, several hundred thousand internally displaced persons have been forced to flee their homes and live under difficult conditions in zones of continuing armed conflict or in government-controlled relocation sites. While some of these people have achieved a level of stability in their new settlements, many have yet to find durable solutions to their plight.

Pockets of relative stability: the KNU cease-fire

After an aborted series of meetings in the mid 1990s, cease-fire negotiations between the SPDC and the KNU began in December 2003 with the announcement of a 'gentleman's agreement' to cease fighting. Although substantial talks began in January the next year, the purge of the relatively progressive prime minister (and military intelligence chief) General Khin Nyunt, in October 2004, presented a serious set-back to the peace process. If the provisional KNU–SPDC cease-fire can be consolidated, it could yet deliver a substantial improvement in the human rights situation on the ground, creating the space in which local and international organisations can begin to address the urgent needs of a war-ravaged population. Since early 2006, however, the *Tatmadaw* has launched major operations against the civilian population and a diminished KNU insurgency across northern Karen State.

Between February and December 2006, some 25,000 people were displaced by *tatmadaw* attacks on villages in northwest Karen State (Toungoo and northern Nyaunglebin Districts, and parts of Papun District—KNLA second, third and fifth brigades). In addition, since April, dozens of villages have received orders from the *Tatmadaw* to relocate to new settlements in areas more firmly under government control.

Recent *tatmadaw* offensives in Karen areas (especially in Papun District) seem designed in part to gain control of previously contested areas, in order to undertake major infrastructure developments, such as the construction of a series of hydroelectric dams on the Salween River. If built—at an estimated cost of more than $5 billion—the dams will flood an estimated 995 square kilometres of forest. In November 2004, a coalition of Karen non-governmental agencies reported that three-quarters of the 85 villages in the vicinity of the planned dam sites had been forcibly relocated since 1995, displacing tens of thousands of civilians. Thus, the fundamental causes of displacement for many new internally displaced persons in Karen (and Karenni and Shan) areas are related to major new development projects. The typology presented above represents a continuum of (overlapping) 'ideal types', rather than discrete categories of forced migration.

These disturbing developments notwithstanding, since the provisional KNU cease-fire, the situation in other Karen areas has begun to stabilise. In parts of Tenasserim Division, and across much of central and southern Karen State, there is less fighting and somewhat fewer human rights violations than before. In October 2004, the TBBC reported that 'more than half [57 per cent] of internally displaced households [had] been forced to work without compensation and…extorted cash or property within the last year'. By October 2005, these numbers had dropped to one-third of those surveyed having paid arbitrary taxes or been subject to forced labour in the past year. In general, therefore, human rights abuses had declined since 2004—at least for those living beyond zones of continuing armed conflict.

Changing patterns of displacement and rehabilitation

Since 2004 and the provisional KNU cease-fire, large numbers of Type One internally displaced persons in central and southern Karen areas have begun to return 'spontaneously' from hiding places in the jungle (and from relocation sites, and some refugee camps in Thailand) to build more permanent (wooden) houses and grow crops other than swidden rice. Especially in central Karen State, many internally displaced persons have moved from cease-fire zones into relatively more secure villages and peri-urban areas, influenced by the government and armed groups (the KNU controls no cease-fire zones).

As noted above, on receiving relocation orders or becoming subject to other forced migration pressures, some people enter relocation sites while others go into hiding in the jungle, move to other villages (including in cease-fire zones) and/or urban and peri-urban areas. Most relocation sites seem to be disbanded within a few years of their establishment, as the authorities turn a blind eye to forcibly relocated communities' efforts to return to their original land or resettle elsewhere. In many cases, however, conditions in relocation sites return to normalcy (by the standards of rural Burma) over time, as people rebuild their communities in the new location, often in partnership with community based organisations (CBOs) and local non-governmental organisations (NGOs). In such cases, residents could prefer life in the new village to the uncertainties of return or resettlement elsewhere and the possibility of being subject to a new round of displacement in the future. Such rehabilitated relocation sites could offer better health and education services and access to markets than the remote village that people were originally forced to vacate.

In such cases—those in which displaced people come to find the new settlement preferable to their original villages—the label 'relocation site' is not particularly helpful. Certainly, people's vulnerabilities and needs and the options for outside intervention will be different to those of people in classic relocation sites. Thus the importance of a community-based approach to needs analysis, which takes account of local responses to displacement. These distinctions also indicate that for many displaced people, rehabilitation *in situ* (a form of spontaneous rehabilitation) will be a preferred durable solution. These comments notwithstanding, many villagers remain ready to flee at short notice, and still often spend a night under the stars if a *tatmadaw* patrol approaches the village. Furthermore, many armed conflict-affected (especially border) areas remain heavily mined, with important implications for any future refugee/internally displaced person repatriation or rehabilitation activities.

Type One: responses and impacts

Type One forced migrants' vulnerabilities and consequent needs vary according to their response to displacement pressures. For example, given orders to relocate, villagers could adopt one or more of the following strategies (plus the increasingly difficult and dangerous option of seeking refuge in a neighbouring country)

- hide in or close to zones affected by continuing armed conflict and forced relocation (with the hope of returning home, but often remaining mobile for years)
- move to a relocation site
- enter a cease-fire area
- move to relatively more secure villages, towns or peri-urban areas, including behind the front lines in war zones, in cease-fire zones and in government-controlled locations.

In many cases, civilians from the same community and subject to the same migration pressure (for example, a relocation order) will adopt a variety of different responses. This is often the case within an individual family: elderly people could attempt to stay at home, adults will go into hiding in the jungle, enter a relocation site or seek new livelihood options in relatively more secure and stable villages, towns or urban areas, while some children could be sent to join relatives in town. A displaced family or individual is more likely to adopt a life in hiding, in a zone of continuing armed conflict, if they have some form of pre-established relationship with an armed opposition group—such as relatives already living in insurgent-controlled areas, or family or friends in the KNU (for example). Similarly, Type One internally displaced persons will tend to enter a cease-fire area, or relocation site, if they have non-threatening relations with the relevant cease-fire group or state authorities.

Type Two forced migration

Type Two forced migration—that is, state–society conflict-induced displacement—is well illustrated by the situation in cease-fire zones in Kachin and Mon States, where populations are disrupted by military occupation and development activities. Type Three forced migration—livelihood vulnerability-induced displacement—is discussed here with particular reference to the impact of opium-growing bans in Kokang Special Region One.

Unlike Type One forced migration, Type Two typically comes about after armed conflict has ceased. In Kachin and Mon States, since the agreement to cease-fires between the government and most insurgent groups in the mid 1990s, armed conflict-induced displacement has come to an end (with the exception of some parts of southern Mon State). Other patterns of forced displacement, however, continue. In the past decade, local communities have lost large amounts of land (and associated livelihoods) to confiscation by the *Tatmadaw*—often in the context of its self-support policy—and by local authorities and business groups, including in the context of development projects and due to unsustainable natural resource extraction. Furthermore, civilians in these areas continue to be subjected to forced labour and other human rights abuses.

Nevertheless, the Kachin Independence Organisation (KIO), New Mon State Party (NMSP) and some other cease-fire groups and their local civil society partners have implemented a range of resettlement, rehabilitation and development programs, despite limited human and financial resources. More could have been achieved with greater government and international financial and capacity-building support. There has, however, been a peace dividend in Kachin and Mon States, and the post cease-fire re-emergence of civil society networks is encouraging.

The government's attitude towards the Kachin and other cease-fire areas has generally been one of neglect or active obstruction. Kachin leaders claim that the SPDC wants to keep their area underdeveloped and undermine the KIO's standing within Kachin communities. Several other negative developments present worrying precedents in the context of a KNU cease-fire. Although there have been no 'four cuts'-type forced relocations in Kachin State since 1983, communities continue to lose their land. Some eleven *tatmadaw* battalions in Bhamo District in southern Kachin State, for example, had by 2004 reportedly confiscated 3–4,000 acres of land. Thousands of people have been displaced by large-scale jade-mining around Phakant, in western Kachin State. Increased post cease-fire logging and gold-mining activities have also brought environmental damage to several areas. Finally, the State's leasing of land to private companies often involves land confiscation, as does development-induced displacement—for example road, bridge and airport construction in the state capital of Myitkyina.

The Mon State case illustrates similar themes. Between 1993 and 1996—and especially after the 1995 NMSP cease-fire—about 10,000 Mon refugees were forced up to and across the border by Thai authorities. Mon refugees were repatriated to NMSP-controlled cease-fire zones with assistance from international NGOs; the United Nations High Commissioner for Refugees (UNHCR) offered neither protection nor assistance. Some refugees returned home, but most remained in limbo, in camp-like conditions just inside the Burma border, with only limited access to agricultural land. Those Mon who did return home continue to face chronic livelihood and food security problems and remain partially dependent on decreasing humanitarian aid. Meanwhile, as a consequence of continuing human rights abuses (and renewed outbreaks of insurgency) in Mon State, newly displaced villagers continue to seek refuge in the Mon cease-fire zones and refugee resettlement sites.

As in Kachin State, the most serious post cease-fire problems in Mon State relate to housing, land and property rights: since 1998, more than 11,000 acres of farmland have been confiscated by the *Tatmadaw*.

Adding insult to injury, farmers have sometimes been forced to work on the confiscated land, building barracks and farming on behalf of the *Tatmadaw* (Human Rights Foundation of Monland 2003). The building of infrastructure on confiscated land using forced labour has resulted in development-induced displacement.

All of these factors have been causes of continued forced migration since the cease-fire, although the reasons for displacement have changed. In many cases, the abuses outlined above undermine villagers' livelihoods so severely that they have little choice but to migrate.

There have, however, been positive developments in Kachin and Mon States in the past decade. Cessation of armed conflict has generally improved conditions of human security—at least in areas where the cease-fire has held. These truces have brought new opportunities to develop local agriculture and for travel and local trade; they have also created the political and military space for the expansion of civil society networks.

Type Three forced migration

Type Three (livelihood insecurity-induced) internal migration is more widespread than the more acute types of forced migration in Burma (types One and Two). Type Three migrants are not ordered or physically compelled to move by the use or threat of force. They can, however, be described as forced migrants in that they generally have little or no meaningful choice other than to move. This type of movement could be referred to as 'distress migration' or 'migration for survival'. Type Three forced migrants constitute a particularly vulnerable subgroup of the larger economic migrant population.

After a 1989 cease-fire with the government, the Kokang cease-fire zone underwent an economic boom as a result of increased opium harvests and heroin-refining activities. The Myanmar National Democratic Alliance Army (MNDAA) ceasefire group and regional *tatmadaw* commanders benefitted financially, even if most were not involved directly. The local Kokang and other ethnic minority communities also benefitted somewhat from the opium boom of the 1990s. Most villagers, however, remained very poor and grew opium poppies only to fill a rice deficit caused by the poor growing conditions for paddy in the steep Kokang hills.

In 1997, the MNDAA announced a ban on growing and processing opium. This was brought on by a combination of government and international (Chinese and United Nations) pressure—and the example of drugs-free development in neighbouring China. By 2002, the ban had been implemented across much of Kokang. It resulted in farmers' incomes dropping by, on average, 70 per cent, leading to extreme livelihood and human security shocks. Plummeting standards of living have led to health and nutrition crises and rising high school drop-out rates, as well as serious environmental impacts. The humanitarian crisis caused extensive, mostly non-voluntary out-migration to China and the Wa cease-fire areas—where villagers could continue to grow poppies, for a while at least. One-third of the population (estimated at 180,000) reportedly migrated from Special Region One (ceasefire zone) in 2003 after the opium ban.

The hillsides of Kokang are ideally suited to poppy cultivation. Even for those who own land, however, it seems unlikely that this terrain could support more than six to nine months of rice needs. The future looks particularly bleak for the 20–30 per cent of the cease-fire zone population who are not ethnic Kokang. Few Palaung, Miao Tser and Lisu villagers own their own fields, having worked previously as day labourers for Kokang villagers and/or Chinese and other opium entrepreneurs. These communities are finding it particularly difficult to switch to alternative livelihoods. In the event of humanitarian aid being withdrawn, a significant proportion of the population will have little choice but to leave Kokang.

If lessons are not learned from Kokang, the impacts of the opium ban—and resulting vulnerabilities—are likely to be reproduced in zones controlled by the United Wa State Party and elsewhere. One by-product of opium-eradication policies in Wa areas has already been the forcible relocation of some 65,000 villagers from opium-growing areas in the northern Wa sub-state (Jelsma, Kramer and Vervest 2005).

Humanitarian protection

Humanitarian, development and political actors' abilities to understand the complexities of forced migration in Burma are particularly important given the evidence from Kachin and Mon States that conflict and displacement did not come to an end with the cessation of insurgency. Kachin and Mon States also indicate the range of projects than can be implemented by local authorities (cease-fire groups) and civil society (CBOs and local NGOs) in the context of less than ideal cease-fires in previously armed conflict-affected areas. More might be achieved with greater support from the government and international agencies.

As noted, since a provisional cease-fire was agreed to between the government and the KNU, the situation in some Karen areas has begun to stabilise. Across parts of lower and western Karen State, there is less fighting and fewer acute human rights violations than before. (Civilians are still, however, subject to a range of abuses, including new problems similar to those experienced post cease-fire in Kachin and Mon States.) These developments raise the subject of displaced people's rehabilitation, including issues of resettlement and return. The primary concern relates to durable solutions—including aid intervention that links relief and development.

An important set of issues to be resolved relates to the rights of refugees and internally displaced persons to return to and recover their original homes, lands and properties. In June 2005, the UN Sub-Commission on the Promotion and Protection of Human Rights endorsed a set of *Principles on Housing and Property Restitution for Refugees and Other Displaced Persons*, which reflected international human rights and humanitarian law (UNHCR 2005; *Forced Migration Review* 2006). The 'Pinheiro Principles'[1] constitute the first consolidated global standard on the housing, land and property rights of the displaced.

Due to the prevalence of refugee-oriented mind-sets, humanitarian and political strategists often assume that all displaced persons want to go home (the equivalent of refugee repatriation, but without the legal protection element). The primary research, however, cautions against such assumptions: at least some Type One and other forced migrants could prefer to remain *in situ*—especially if their concerns for physical security are addressed adequately. Other displaced persons will want to resettle elsewhere—either returning home or moving to a new location—especially if sustainable solutions are found to long-running armed and state–society conflicts in Burma.

The durable solution of local integration could allow internally displaced persons to escape cycles of displacement and begin to rebuild their lives. Whether they want to stay in their present settlement or return to a previous home will in part depend on their current state of livelihood and human security *in situ*—for example, whether they have found at least semi-durable solutions to their plight. Another important factor will be their knowledge of what has happened to their old homes, land and other property, and whether these have since been occupied—by the State or the *Tatmadaw* (or other armed group), by private commercial interests (often linked to state or para-state agencies) or by other civilians (secondary occupants—quite possibly, other internally displaced persons). As in refugee repatriation, the principle of informed voluntariness should be central to any decisions regarding solutions to internal displacement in Burma.

The protection of internationally agreed rights is first and foremost the responsibility of states. Not all states, however, are signatory to all aspects of international law. The Burmese government has not ratified the instruments of the UNHCR or the International Covenant on Economic, Social and Cultural Rights. In cases such as Burma, where the State is unwilling or unable to protect its citizens, this responsibility falls on the international community. Some international agencies (for example, the UNHCR, UNICEF and the International Committee of the Red Cross) are tasked specifically with protecting certain rights, or categories of people. In addition, the United Nations has a mandate to protect and promote human rights.

At its broadest, the notion of humanitarian protection includes securing access to the right to life (for example, physical security and the rights to shelter, food and water). This could take the form of material aid (substitution mode) supplied directly to the target population—for instance, the distribution of rice by the World Food Program to communities that have suffered as a result of opium bans in Shan State. Humanitarian actors could also work in partnership with state or non-state actors to deliver goods and services. For example, UNICEF supports the SPDC ministries of health and education by providing training to staff and funding the acquisition and distribution of medicines (including vaccination campaigns) and teaching materials. Other international donors support local Burmese NGOs and CBOs to provide a range of services to displaced populations in Burma—often in conflict-affected areas that are beyond the reach of international agencies.

Humanitarian assistance alone tends to be responsive or remedial in nature. This mode of intervention is often, however, insufficient to alleviate suffering and protect human dignity, because it does not address the underlying causes of distress. The concept of protection implies prevention, which in turn draws attention to the reasons for deprivation. It is often necessary to address the actors and structures that cause violence and suffering; however, in a constrained working environment such as Burma, it is easier to focus on service delivery and relief activities than on more politically challenging issues such as protection. There is a danger that power-holders (including especially the State) could withdraw access to vulnerable populations. This access is necessary in order to deliver assistance, should the humanitarian actor seek to engage power-holders on these issues.

Therefore, one of the greatest challenges facing international agencies in Burma is how to achieve a balance between short and longer-term assistance interventions, while keeping a focus on protection concerns. 'Assistance versus protection' is not a zero-sum game: where assistance access is possible, often forms of protection can also be provided. By employing a range of strategies, including supporting the practices of affected communities, it is often possible to address protection concerns in the process of meeting other basic needs.

Advocacy

According to Slim and Bonwick

> ...advocacy is a core area of protective practice for both humanitarian and human-rights agencies. It is about convincing decision-makers to change...It encompasses everything from persuading the village chief to allocate land to displaced families to influencing a senior General on the conduct of his army (2005:84).

Humanitarian advocacy aims to protect civilians from—or alleviate the impacts of—abuse. Such action falls under three broad modes: denunciation, mobilisation and persuasion ('responsibalisation').

Some agencies—primarily human rights-oriented groups based outside the country—denounce the violation of basic rights involved in forced displacement and call for fundamental changes in Burma, or at least radically improved behaviour on the part of the State and armed groups. In most cases, their recommendations are very general, with few attempts to seriously consider how suggestions might be achieved in practice. In general, those who are affected most by armed conflict and cease-fires have the least ability to influence such public advocacy agendas—they are denied a 'voice'—in comparison with relatively well-educated urban and political élites and are rarely consulted in setting advocacy goals and messages regarding their plight.

Organisations working in government-controlled Burma cannot afford to be as bold in their advocacy roles as those in Thailand and overseas; however, the presence of humanitarian personnel in conflict-affected areas can help to create a 'humanitarian space' in which to engage in behind-the-scenes advocacy. A consciously adopted and visible protective presence could constrain local power-holders' opportunities for abuse, because authorities worry that information regarding violations will be communicated to the international community and/or because the presence of a witness 'shames' them into adopting better behaviour. This is a persuasive mode of advocacy.

This is an area in which UN agencies and the International Committee of the Red Cross (ICRC) have made some progress in the past few years. Confidential advocacy with national, state and local authorities has helped to build a more protective environment, especially in the fields of harm reduction and HIV/AIDS issues, trafficking and child rights. Examples include the establishment of an interagency–government committee to stop recruitment and facilitate demobilisation of child soldiers and ICRC and UNHCR training programs for police and other government employees. The advocacy activities of the ICRC—including its confidential referral of cases of human rights abuse to the authorities—also gives some leverage to progressive elements within the government and state agencies, who wish to establish better practice in their fields.

Some UN agencies have specific, and therefore rather restricted, protection mandates. For example, UNICEF has made progress in a number of protection sectors, with the government recognising child protection concerns and implementing new initiatives. Indeed, UNICEF (2005) has been able to persuade the government that more strategic approaches are required to 'reach the unreached'—for example, focusing in the education sector on 'the most vulnerable, including poor, minority and out-of-school children, children living in remote areas, and children from migrant and mobile populations'. The ICRC's protective presence in areas of continuing armed conflict has also been quite effective, at least until it had its access significantly curtailed in 1995.

Some civil society groups with programs inside Burma have also mobilised agencies operating in persuasive or denunciation modes. For example, CBOs in rural areas could pass on human rights information to their local and international counterparts in Yangon or Thailand. There is evidence that the existence of such protection and advocacy networks has served to reduce the incidence of human rights abuses in some parts of Karen and Karenni States. Most international (and especially UN) agencies inside Burma, however, demonstrate only limited awareness of protection issues, and undertake minimal advocacy activities on behalf of displaced persons. As a senior UN officer explained to the author,

'In general, and with some important exceptions, there is a lack of a "culture of protection" within the UN, especially at the field level.'

In December 2005, the UN Inter-Agency Standing Committee assigned to the UNHCR primary responsibility for leading a cluster of agencies in coordinating assistance to and the protection of the estimated 20–25 million internally displaced persons world-wide (*Forced Migration Review* 2006). It is hoped that in the next two years adoption of the new cluster will prompt international agencies to address gaps in responses to internally displaced person crises in Burma.

Restrictions on humanitarian space

The ability of local and international agencies to address Burma's protracted and interrelated displacement crises is in large part determined by the amount and quality of political and humanitarian 'space' available. The period from November 2003 to September 2004 was one of rapidly opening humanitarian space in Burma. In part, the authorities' willingness to allow international access to previously out-of-bounds areas was a response to increased pressure after the 'Depayin Massacre' of 30 May 2003.

Since October 2004 and the demise of Khin Nyunt and colleagues, along with their relatively progressive ideology, the extent and quality of political and humanitarian space in Burma has declined. For humanitarian agencies, this constriction is reflected in a set of draft *Guidelines for UN Agencies, International Organizations and NGO/ INGOs on Cooperation Program in Myanmar* produced by the Ministry of National Planning and Economic Development office in February 2006. Some of its more worrying proposals include that state officials should accompany UN and international NGO staff on all field trips; the proposed supervisory roles are to be played by central, state-divisional and township coordinating committees (including roles for the Union Solidarity Development Association and various government-operated NGOs); and the government plans to vet all new Burmese staff of the United Nations and international NGOs.

It seems likely that, should these regulations be implemented systematically, some international agencies will withdraw from the country. Already the Global Fund for HIV/AIDS, Malaria and Tuberculosis has ceased operations in Burma—although in this case there were additional, politically driven considerations behind the decision to leave. In February 2006, Médecins sans Frontières (MSF) also withdrew from Burma, claiming that increased government restrictions imposed since 2005 had made its operations in Mon and Karen States untenable. As MSF avoids working with local state structures, and thus does little to build local capacities, it was ill-prepared to operate in an increasingly constricted humanitarian environment.

A further consequence of the restrictive operating environment in Burma is that most international agencies have very limited access to the upper echelons of the military government, and are unable to engage in policy dialogue with, or communicate advocacy messages to, the regime. Nevertheless, given these increasingly bleak and repressive conditions, the United Nations in particular has a special responsibility to advocate and act on behalf of the most vulnerable, conflict-affected populations.

Conclusions

This chapter has described aspects of forced migration in Burma that are under-researched, including the phenomenon of serial displacement, and has proposed a three-part typology. Many internally displaced persons and others move repeatedly, sometimes for a combination of reasons; others have been displaced for some time and have found at least semi-durable solutions to their plight; many are living mixed with communities who are not—or have not recently been—displaced. Forced migrants' needs can be assessed and appropriate interventions planned only if the full complexity of displacement situations in Burma is understood. Humanitarian (and political) actors should therefore respect and respond to the voices and agency of forced migrants and enrol their participation in all aspects of program planning and implementation. In most cases, forced migrants and communities threatened by displacement have special protection vulnerabilities related to the causes of migration (especially armed and state–society conflict). These concerns link humanitarian needs to explicitly political issues. Ultimately, substantial and sustained protection from forced migration, as well as the rehabilitation of displaced populations and reconstruction of communities, depends on resolutions to the conflicts that cause displacement in Burma. Unfortunately, efforts at conflict resolution have thus far met with only limited success.

Notes

1   The 'Pinheiro Principles' are the United Nations' Principles on Housing and Property Restitution for Refugees and Displaced Persons, which were prepared by the United Nation's Special Rapporteur, Professor Paolo Sergio Pinheiro, and adopted by the United Nations in August 2005. They have been published by the Centre on Housing Rights and Evictions (COHRE).

References

Amnesty International, 2002. *Myanmar: lack of security in counter-insurgency areas*, Amnesty International, London.

——, 2004. *Myanmar—the Rohingya minority: fundamental rights denied*, Amnesty International, London.

——, 2005. *Thailand: the plight of Burmese migrant workers*, Amnesty International, London.

Burma Ethnic Research Group and Friedrich Naumann Foundation, 1998a. *Forgotten Victims of a Hidden War: internally displaced Karen in Burma*, Burma Ethnic Research Group, Chiang Mai. Available from http://www.ibiblio.org/obl/docs3/Berg-Forgotten_Victims.pdf

Bamforth, V. Lanjouw, S. and Mortimer, B., 2000. *Conflict and Displacement in Karenni: the need for considered responses*, report prepared for the Burma Ethnic Research Group, Chiang Mai.

Burma Issues, 2003. *After the 1997 Offensives: the Burma Army's relocation program Kamoethway Area, Tenasserim Division*, The Peace Way Foundation, Bangkok.

Callahan, M., 2003. *Making Enemies: war and state building in Burma*, Cornell University Press, Ithaca, New York.

Caverzasio, S.G. (ed.), 2001. *Strengthening Protection in War: a search for professional standards*, International Committee of the Red Cross, Geneva.

Cusano, C., 2001. 'Burma: displaced Karens: "Like water on the *Khu Leaf*"', in M. Vincent and B.R. Sorensen (eds), *Caught Between Borders: response strategies of the internally displaced*, Pluto Press for Norwegian Refugee Council, Oslo.

Free Burma Rangers, 2006. *Map of Burma Army Attacks in Northern Karen State*, 28 June.

Grundy-Warr, C. and Wong, E.S.Y., 2002. 'Geographies of displacement: the Karenni and the Shan across the Myanmar–Thailand border', *Singapore Journal of Topical Geography*, 23(1):93–122.

Heidel, B., 2006. *The Growth of Civil Society in Myanmar*, Books for Change, Bangalore.

Heppner, K., 2005. *Sovereignty, survival and resistance: contending perspectives on Karen internal displacement in Burma*, Karen Human Rights Group Working Paper, Karen Human Rights Group, Bangkok.

Humanitarian Affairs Research Project, 2003. *Running the Gauntlet: the impact of internal displacement in Southern Shan State*, Asian Regional Centre for Migration, Chulalongkorn University, Bangkok.

Human Rights Foundation of Monland, 2003. *No Land to Farm: a comprehensive report on land, real estate and properties confiscation in Mon's Area, Burma (1998–2003)*, Human Rights Foundation of Monland, Bangkok.

Human Rights Watch, 2005. '"They Came and Destroyed Our Village Again": the plight of internally displaced persons in Karen State', Human Rights Watch Report, 17(4), January. Available from: http://hrw.org/reports/2005/burma0605/burma0605.pdf.

Jelsma, M., Kramer, T. and Vervest, P. (eds.), 2005. *Trouble in the Triangle: opium and conflict in Burma*, Silkworm Books, Chiang Mai.

Karen Rivers Watch, 2004. *Damming at Gunpoint: tatmadaw atrocities pave the way for Salween Dams in Karen State*, Karen Rivers Watch, Chiang Mai.

Lang, H., 2002. *Fear and Sanctuary: Burmese refugees in Thailand*, Cornell, Ithaca.

Mon Language Literacy Training Course, 2005. (unpublished) '2005 report', The Mon Literacy Training Course Organizing Committee, August.

Shan Human Rights Foundation, 2003. *Charting the Exodus From Shan State*, Shan Human Rights Foundation, Chiang Mai.

Sherman, J., 2003. 'Burma: lessons from the ceasefires', in J. Ballentine and K. Ballentine (eds), *The Political Economy of Armed Conflict: beyond greed and grievance*, Rienner, Boulder.

Slim, H. and Bonwick, A., 2005. *Protection: a guide for humanitarian agencies*, Active Learning Network for Accountability and Participation and Overseas Development Institute, London.

Smith, M., 1999. *Burma: insurgency and the politics of ethnicity*, Zed Books, London.

South, A., 1994. 'Political Transition in Burma: a new model for democratisation', *Contemporary Southeast Asia*, Institute of Southeast Asian Studies, Singapore.

——, 2005. *Mon Nationalism and Civil War in Burma: the golden sheldrake*, Routledge, London.

Taylor, R., 1985. 'Government responses to armed communist and separatist movements: Burma', in C. Jeshurun (ed.), *Governments and Rebellions in Southeast Asia*, Institute of Southeast Asian Studies, Singapore.

——, 1987. *The State of Burma*, Hurst, London.

Thailand Burma Border Consortium, 2004. *Internal Displacement and Vulnerability in Eastern Burma*, Thailand Burma Border Consortium, Bangkok.

——, 2005a. *Program Report*, Thailand Burma Border Consortium, Bangkok.

——, 2005b. *Internal Displacement and Protection in Eastern Burma*, Thailand Burma Border Consortium, Bangkok.

——, 2006. *Burmese Border Refugee Sites With Population Figures 2005*, Thailand Burma Border Consortium, Bangkok.

Thant Myint-U., 2001. *The Making of Modern Burma*, Cambridge University Press, Cambridge.

UNHCR, 1998. *The Guiding Principles on Internal Displacement*, E/CN.4/1998/53/Add.2, United Nations High Commissioner for Refugees, Geneva.

——, 2005. *Principles on Housing and Property Restitution for Refugees and Displaced Persons*, E/CN.4/Sub.2/2005/17, United Nations High Commissioner for Refugees.

UNICEF, 2005. *The Situation of Women and Children in Myanmar*, Country Program Document, Yangon.

Acknowledgments

Research was conducted during consultancies for the Thailand Burma Border Consortium (2002), International Crisis Group (2003), Human Rights Watch (2004–05), United Nations Development Programme (2005) and with a grant from the John D. and Catherine T. MacArthur Foundation (2003–04). Special thanks for help with research and writing to Monique Skidmore, Julie Belanger, Alan Smith, Martin Smith, and to many friends and colleagues in and from Burma. A version of this chapter was published by the Refugee Studies Centre (Oxford University) in February 2007 (RSC Working Paper No. 39).

5 Foreign policy as a political tool: Myanmar 2003–2006

Trevor Wilson

In the last quarter of 2004, observers were uncertain how the new leadership would handle Myanmar's international relations, notwithstanding the continuity at the top of the regime. Spokesmen for the new leadership initially were at pains to reiterate their continuing commitment to Myanmar's opening up to the world. Early statements were deliberately cast in reassuring terms for Myanmar's most important neighbours—although these statements were very general. The key statement was by the State Peace and Development Council (SPDC) spokesman Lieutenant-General Thein Sein, who gave a commitment that the national reconciliation 'road-map' would continue after the change of prime minister, because this was state policy and 'not the concern of a single individual'.[1] A week after the dismissal of General Khin Nyunt as Prime Minister, Senior General Than Shwe was on a state visit to India, while within two months the new Prime Minister, General Soe Win, began visiting Association of Southeast Asian Nations (ASEAN) neighbours. All of these occasions were used to demonstrate the regime's continued interest in foreign investment, tourism and, above all, 'friendly' relations with its neighbours. At that stage, at least, there was no trying to turn back the clock.

Initially, the new SPDC leadership did not rush either to cancel approvals given to international assistance programs or to stop international non-government organisations (NGOs) operating in their various humanitarian and capacity-building activities. It was tempting at this point to hope that it might be business as usual for the international community's operations in Myanmar, as most donors of international assistance continued their programs and patiently sought to resume more sensitive projects. But subsequent decisions by the authorities reveal decidedly more negative patterns and trends. Cooperation with United Nations agencies was particularly fraught, as it gradually became clear that the new leadership would continue to refuse access to UN Special Envoy, Razali Ismail, and UN Special Rapporteur for Human Rights, Sergio Paolo Pinheiro, each of whom had served for several years with some success and shown considerable understanding of Myanmar's position. While the leadership did not repudiate all cooperation with the United Nations after 2004, it was prepared to go to the brink in its relationship with the International Labour Organization (ILO).

As the post-2004 Myanmar leadership refined its handling of its overall policy approaches, it became evident that foreign policy did not enjoy the same priority that it had under Khin Nyunt, and was more than ever before subordinated to domestic military policy objectives and less influenced by 'professional' diplomatic considerations. While Prime Minister, General Soe Win, and the new Foreign Minister, Nyan Win (a former army officer), took on the responsibilities for representing Myanmar at international meetings, Vice-Senior General Maung Aye was believed to be playing a more active role behind the scenes, and a new conservative voice was the new Labour Minister, U Thaung, another retired military officer and former Ambassador to Washington (who was also Minister for Science and Technology). As a result, foreign policy was more reactive and defensive than before, partly reflecting the new leadership's lack of international experience, while the increased military domination of foreign policy made it more introverted, more security conscious and less cooperative than ever before. These tendencies were only partly a product of the regime's own inward-looking character; they were also a response to the wider international environment of aggressive US unilateralism (in Iraq and elsewhere), to the tightening of selective bilateral sanctions against Myanmar by some countries and to the world-wide fixation with the threat of terrorism and 'rogue states' generally. Moreover, the previous more outward-looking international policies were being questioned, and sometimes jettisoned, merely because they originated with dismissed Prime Minister, General Khin Nyunt.

While some observers saw the new leadership as more isolationist, in fact it maintained a high level of activity in its relations with its neighbours between 2004 and 2006. Myanmar's greatest diplomatic triumph in this period was its inclusion in the Asia-Europe Summit meeting in Hanoi in October 2004, even though it participated at foreign-minister rather than head-of-state level. The Myanmar government could feel pleased with this victory when most other trends were not so favourable. Between 2004 and 2006, however, the new leadership's inflexible stance against its domestic political opponents generated growing discomfort internationally, even among some of the regime's most trusted 'friends' in its own Asian region. Moreover, the Myanmar leadership's more negative attitude towards the United Nations generated more intense international questioning of its readiness to cooperate with the international community.

Interaction with ASEAN

The main arena for Myanmar's international interaction remained its relationship with ASEAN and its collective and bilateral associations with its fellow ASEAN members. Significantly, one of the major changes in Myanmar's foreign relations in the three years from 2003 to 2006 was the far greater readiness of ASEAN to criticise and seek to influence Myanmar on its domestic political policies. Since 1997, ASEAN had been compliant and publicly uncomplaining about Myanmar, but this changed after the detention of Aung San Suu Kyi in May 2003. When the Myanmar government remained resistant to outside requests to announce a time frame for political reform, ASEAN became uncharacteristically vocal in expressing its concerns. Another reason for ASEAN's discomfort was the increased international attention on it because Myanmar was to assume the chair of the association in 2007. This represented a major foreign policy dilemma for Myanmar, and generated tension between Myanmar and its fellow ASEAN members for much of the period after 2004. In addition, after a campaign by expatriate Burmese political activists, politicians from other ASEAN countries formed the anti-SPDC ASEAN Inter-Parliamentary Union Myanmar Caucus, over which ASEAN governments had little control.[2] ASEAN countries could also not ignore Washington's reluctance to include Myanmar in any ASEAN activities with which it was associated, which complicated the conduct of ASEAN-plus activities, even if US diplomacy was not always very adroit. Nevertheless, after a long tussle—much of it, unusually for ASEAN, conducted through the media—Myanmar finally announced in August 2005 that it would not insist on its turn as chair, offering the implausible excuse that the government would be preoccupied with its national reconciliation process.

By choosing to step aside as ASEAN chair, the Myanmar government failed in a cherished strategic policy objective, evident in its own statements in the years of Khin Nyunt's ascendancy in foreign policy, and in this case maintained after his departure. These statements had made clear the SPDC's keenness to demonstrate its international credentials and legitimacy by hosting the ASEAN Summit when its turn came up in 2007, and the Myanmar government initiated specific preparations for hosting the summit in many areas. Yet, while stepping down as ASEAN chair was a loss of face for Myanmar and humiliating for Myanmar's leaders, it was preferable to submitting to external pressure over the vital issue of political reform. Moreover, Myanmar was able to make a virtue of its decision tactically, as it helped Myanmar's fellow members of ASEAN extricate themselves from a difficult political situation.

Ultimately, stepping down probably achieved no more than buying more time for Myanmar and ASEAN. It did not foreshadow any change in Myanmar's approach, and in itself did not contain the ingredients for a compromise between Myanmar and ASEAN. In its collective responses to Myanmar's intransigence, ASEAN had for some time been a prisoner of its own traditional policy of non-intervention in the internal affairs of other members, a position reiterated publicly by ASEAN Secretary-General, Ong Keng Yong, as late as June 2006.[3] Although Myanmar's diplomatic representatives worked hard to defuse this issue within ASEAN, they met decreasing success after 2003, and senior Myanmar spokesmen became annoyed by this perceived pressure from, and double standards being applied by, ASEAN.[4]

New signs of ASEAN resolve emerged in early 2006 when Malaysian Foreign Minister, Syed Hamid Albar, sought to visit Myanmar with an announced tougher mandate from ASEAN, and became the first ASEAN leader to seek to meet Aung San Suu Kyi. The Myanmar government unwisely irritated Albar, first by keeping him waiting for two months to make his visit and then by not allowing him to meet Aung San Suu Kyi. Despite Albar's failure to meet Aung San Suu Kyi, he continued to express public concern about the lack of progress towards reconciliation. The June 2006 SPDC decision to extend Aung San Suu Kyi's house arrest for another year was thus in part a direct rebuff to ASEAN and, if anything, produced a further toughening of ASEAN's position. Albar's July 2006 public response, that ASEAN 'could not defend Myanmar' (Albar 2006), merely underlined the clumsiness of the SPDC's handling of ASEAN.

From Myanmar's point of view, a series of visits from the heads of government of all the main ASEAN members during 2004–06— President Susilo Bambang Yudhoyono of Indonesia, Prime Minister Abdullah Badawi of Malaysia, Prime Minister Goh Chok Tong of Singapore and Prime Minister Thaksin Shinawatra of Thailand—went well. Most of these visits were accompanied by the signing of broad agreements on increased bilateral cooperation. The Myanmar government had thus succeeded in reaffirming its bilateral relationships with key

ASEAN countries after the October 2004 political changes, without having to make any noticeable concessions. Although more attention was reportedly paid to Myanmar's political situation during these visits than previously, Myanmar avoided undue public embarrassment over its refusal to release political detainees. As time went on, however, it became increasingly questionable whether Myanmar could maintain indefinitely its position with ASEAN. Rather, it now seemed that a truly satisfactory solution on Myanmar's standing in ASEAN would depend on substantial political changes occurring inside Myanmar.

China, India and Japan

For its part, the new Myanmar leadership can also claim considerable success in consolidating its political and economic relations with China since 2004. A series of high-level visits helped secure important new Chinese investment, valuable concessional loans for infrastructure projects and expanded two-way trade. Myanmar's relationship with China became more important than ever as most other foreign investors and businesses gave up on the country.[5] China's position as a member of the UN Security Council sympathetic to Myanmar's point of view and opposing economic sanctions against Myanmar at a time when this was being actively canvassed in Security Council corridors, illustrates this point. Chinese support almost certainly ruled out any broadening of sanctions.[6] This made China a highly valuable partner and Myanmar's leaders have been openly grateful for China's continued political support.

Yet Myanmar stopped short of total identification with, or subordination to, Chinese interests, and signs of mutual dissatisfaction between the two countries have surfaced more openly since 2003. The most notable recent example of Myanmar standing up to China was the issue of illegal logging in early 2006, when the Myanmar authorities made known their concern about the extent of illegal logging by Chinese entrepreneurs in Myanmar's northern border areas, convincingly documented in a January 2004 report by the environment group Global Witness. Relations with China deteriorated to the point of Myanmar soldiers shooting and killing illegal Chinese loggers, to the apparent annoyance of the Chinese government.[7] In 2003, the Myanmar government did not agree to a Chinese proposal for an integrated shipping route to the Indian Ocean via the Irrawaddy River to Yunnan, because China's request for exemption from customs duty offended the Myanmar leadership's insistence on national sovereignty.[8]

Equally, China's 'embrace' of Myanmar was by no means as open-ended, or its influence as total, as some would argue. Even though China had considerable success after 2003 in gaining access to valuable natural resources in Myanmar, China did not always achieve its goals. China worried about Myanmar's inability to control illicit drug trafficking and requested that this be addressed by the Myanmar authorities, proposing a new bilateral agreement to achieve this. China has been concerned about Myanmar's economic policies, and at times has held off providing loans because of Myanmar's inability to meet repayment schedules. Residual Chinese doubts about the long-term viability of the military regime's policies surfaced more openly in the lead-up to Prime Minister, General Soe Win's, formal visit to China in February 2006, when statements by Chinese President, Hu Jintao, that China 'wanted Myanmar to move towards national reconciliation' hinted clearly at these misgivings. Press reports apparently emanating from Chinese sources in the lead-up to General Soe Win's visit were surprisingly open about China's unhappiness with the situation in Myanmar. When Senior General Than Shwe last visited in February 2003, Vice-Premier, Wu Yi, reportedly told him that China wanted Burmese politics 'to move in a positive direction', as reported in *The Irrawaddy Online Edition*. Reporting from Bangkok more recently, Larry Jagan described Chinese views as including 'reservations concerning the SPDC's lack of progress towards political and economic reform'.[9]

China has long been recognised as potentially playing a key role in Myanmar. Despite the efforts of UN Special Envoy Razali to pursue a dialogue with China on Myanmar, China has not thus far been effectively brought into the UN process of resolving the Myanmar problem, but it could be forced to take a stance in the Security Council deliberations. Although China could be the only country that can influence Myanmar's leadership, it will need to be more overtly engaged in a process to achieve this, and will need to be convinced that it is in its interest to become more proactive rather than pursue the benefits it already receives under the status quo.

Despite much commentary about India's policies towards Myanmar, India remains a second-tier player. By any hard measure of influence or interests—trade, investment, aid, loans, arms sales, gas purchases—India is well behind China, and even Thailand. But certainly the years since 2000 have seen an intensification of India's efforts to develop its relations with Myanmar, through, for example, a series of high-level visits, including by the two heads of state. These visits are largely symbolic, but they illustrate that both sides feel they can develop their relations further. The potential for further development of Myanmar–India relations is undoubtedly great, and there could be fewer inhibitions from Myanmar's point of view than with China. Moreover, India is probably the key to developing better physical infrastructure (roads, rail and ports) in the west of Myanmar.[10] One of the main factors limiting India's influence is that India itself sees its relations with Myanmar essentially in terms of its strategic competition with China. While Myanmar sometimes chooses to take advantage of this competition, it also means that Myanmar's leaders are cynical about India's motives. India's reluctance to confront the problem of Myanmar's political impasse reduces the influence it can exercise, and makes India–Myanmar relations a negative rather than positive factor in terms of encouraging change.

Since 2003, Japan's relations with Myanmar seem to have entered a low point. The Depayin Massacre strengthened the position of pro-democracy supporters in Japan, where political attitudes on Myanmar/Burma are polarised. Japan has not abandoned its policy of 'engagement' or its preference for only limited sanctions, and in June 2006 surprised many by its reluctance to support the inscription of Myanmar on the UN Security Council's agenda. Generally, Japan adopted a much lower profile on Myanmar: between 2004 and 2006, the Japanese government issued fewer public statements on Myanmar than in previous periods, sent fewer official visitors to Myanmar and generally sought a low profile.[11]

While the Japanese government continued to issue protests against the detention of Aung San Suu Kyi, these seemed perfunctory. Japanese official development assistance flows remained significant when assistance from other sources was relatively small.[12] But Japanese official development assistance, which is purportedly for basic humanitarian needs, goes mostly to support Myanmar government activities. Although certain sectors of Japanese politics and business still support developing economic ties, the modest levels of Japanese trade and investment have not improved,[13] reflecting the unattractive commercial environment of Myanmar for Japanese firms. Whatever influence Japanese engagement once had on the Myanmar government was undermined by Japan's weak links with opposition groups and by its obvious desire to retain its links with the government.

United States

Since 2004, the stand-off between Myanmar and key elements of the international community has intensified. The United States retains its leadership of the campaign of outright rejection of military rule in Myanmar. Burma is clearly not of strategic importance to the United States and officially no attempt is being made to engage the military regime. No senior US official has travelled to Rangoon to speak directly to the Myanmar leadership since 2003, reflecting the abandonment of any attempt at direct engagement and its replacement with a ratcheting up of public criticism of Burma.[14] While US determination to place Myanmar on the UN Security Council agenda paid off in September 2006, it is not clear whether there will be effective follow-up action. To date, the Bush administration's strident criticisms of Myanmar, exaggerations in its own reporting of human rights abuses in Myanmar and its obvious subjugation to the partisan Burma lobby in the US Congress, all reduce the potency of US influence. More importantly, there is no evidence of wider US sanctions imposed in the *Burmese Freedom and Democracy Act* of 2003 producing any political concessions by the Myanmar leadership. Rather, the effect of current US policies seems to make the Myanmar regime—which believes it has cooperated substantially with the international community on issues such as narcotics trafficking, religious freedom, money laundering and people trafficking, including by introducing specific legislation—even less compliant.

Although Washington has actively sought international support for its campaign against Burma, it has been only partially successful. The Bush administration's dialogue with Asian countries about Myanmar has not necessarily achieved the support the United States was seeking. Secretary of State, Condoleeza Rice, has only rarely attended high-level ASEAN meetings in ASEAN capitals, where she would meet her ASEAN counterparts (including the Myanmar representatives) on an equal footing, instead mostly choosing to meet them in specially convened meetings outside ASEAN countries. US policy has been partially responsible for ASEAN countries increasing their criticism of Myanmar, but has not persuaded ASEAN to support formal UN Security Council action against Myanmar. In June 2006, even Japan, which the United States at one time claimed was swinging to Washington's point of view,[15] initially opposed the US proposal to inscribe Myanmar on the formal UN Security Council agenda under Chapter VII of the UN Charter. Generally, there is still no US policy with any credible prospect of bringing about its real goal of regime change or providing any realistic 'exit strategy' for the current Myanmar military leadership.

Europe

Myanmar's relations with Europe have long been dominated by attempts by European nations to find collective responses to Myanmar, through the European Union on the one hand and through the Asia-Europe Meeting (ASEM) on the other. The period from 2004 to 2006 witnessed a continuation of these efforts, with the Europeans sometimes succeeding and sometimes failing. Myanmar pursued a rather dogged approach to secure what it believed was its sovereign right to participate in ASEM, and was successful in maintaining ASEAN support for this position.

ASEAN's determination that Myanmar should be included in the ASEM process was the subject of tension between the two sides of the forum for several years, as the Europeans sought to block Myanmar's participation unless the government released Aung San Suu Kyi and moved ahead in its process of political transition. Europe's failure on this partly reflects continuing divisions among European countries on Myanmar, but Europe probably also misjudged its power and leverage on this issue. Yet in terms of substance, the ASEM/Myanmar issue was more symbolic and rhetorical than producing either major consequences or concrete outcomes. ASEM echoed other organisations in criticising Myanmar's non-compliance on issues such as international law enforcement, but it is debatable whether ASEM statements with no specific enforcement make much difference to the behaviour of the Myanmar leadership. Myanmar's presence on the ASEM agenda is, therefore, unlikely to have much impact on Myanmar policy.

With the advantage of hindsight, some Europeans now admit that 'the Asia-Europe partnership as a whole has been held hostage by the Burma/Myanmar issue in 2004' (Pereira 2005). A similar view held by the Asian countries was reflected in remarks by Singapore Prime Minister, Goh Chok Tong, in May 2004: 'the Burma/Myanmar issue disproportionately preoccupied Asia–Europe political exchanges and has become an obstacle to seeking common ground on other strategic issues' (quoted in Pereira 2005). An 'independent' evaluation of ASEM commissioned in early 2006 also singled out the Myanmar issue for the difficulties it caused the organisation, describing it as 'a pressing issue that requires attention', and expressed the vague but optimistic view that 'steps toward a constructive solution to this dilemma could be made at the Helsinki Summit in September 2006'.[16] But after the Helsinki ASEM summit, there was still no sign of a more effective ASEM approach on Myanmar. Nor is there any sign of the Myanmar leadership being influenced by ASEM, now that it has achieved its primary, but essentially limited, goal of achieving recognition of its legitimate standing through its regular participation in ASEM.

On the other hand, EU policy on Myanmar as applied through the European Union's Common Position, became more discriminating in its targeting from 2004 to 2006. EU sanctions became tougher and more selective, although since 2003 the European Union abandoned its attempts via its troika mechanism to engage in meaningful engagement with Myanmar. Yet even in their 'smarter' guise, EU sanctions were isolated measures not implemented uniformly by EU members, and had no greater impact than before.[17] Indeed, some elements of EU sanctions still contained anomalous, and sometimes counter-productive, provisions in relation to investment, freezing assets and visa bans.[18] At the same time, the allowable scope of EU assistance programs to Burma was extended, with assistance for environmental programs allowed under the revised common position from April 2006. Ultimately, EU policies do not attract much attention from the Myanmar leadership, and this refinement of EU policy has had no visible impact on the SPDC. It is hard to disagree with a 2005 assessment that the European Union consistently 'punches below its weight' in Myanmar.[19]

Myanmar's international policies

Overall, foreign policy under the new Myanmar leadership was not only defensive, it lacked innovation. From 2004 to 2006, the SPDC's only new moves were the attempt to rekindle relations with Russia epitomised in Deputy Senior General Maung Aye's visit to Moscow in February 2006, and to resume diplomatic relations with North Korea. On the face of it, neither of these initiatives was likely to change the character of Myanmar's foreign or domestic policies. Russia's main role in Myanmar recently has been as a supplier of important military equipment, probably a key motive behind Maung Aye's visit. Russia's membership of the UN Security Council is also important to the Myanmar leadership, but this support did not require a high-level visit to Moscow. Despite the speculation prompted by Russia's 2002 offer of nuclear research assistance to Myanmar, Russia has otherwise been a minor player in the country. As of October 2006, negotiations on the normalisation of relations between Myanmar and North Korea had still not been finalised, but this seemed to be no impediment to the North Koreans selling conventional arms to Myanmar.[20]

Relations between Myanmar and the UN system and international non-governmental assistance agencies became the source of the greatest foreign policy challenges for all concerned during 2004–06. Since early 2004, the Myanmar government refused to allow either UN Special Envoy, Razali Ismail, or Special Rapporteur on Human Rights, Sergio Paulo Pinheiro, to visit the country in order to pursue their mandates. No reasons were ever given for this, and there is no evidence of either envoy breaching their mandates. The Myanmar leadership, however, appears to consider Razali to be too close to Aung San Suu Kyi and no longer politically neutral. So, when Razali announced in January 2006 that he was stepping down as Special Envoy after waiting almost two years without being allowed to visit, this seemed to represent a set-back for the prospects for international efforts to promote political reconciliation in Myanmar. Although the UN Secretary-General called on the SPDC to resume cooperation with the United Nations, as of late 2006 no successor to Razali had been appointed.

Meanwhile, without the benefit of access to the country, Pinheiro's reports on the human rights situation gradually—and inevitably—became more negative as he relied increasingly on outside reports and as he was frustrated by the lack of cooperation from the SPDC. These reports also became, no doubt, less and less appealing to a sceptical Myanmar government, which was disappointed with Pinheiro after 2002 when it could not persuade him to help refute allegations that rape was being used against ethnic minorities. Professor Pinheiro's term was extended for another year earlier in 2006 even though by then it seemed increasingly unlikely that an effective role for him as Special Rapporteur could be resuscitated.

Similar SPDC suspicions about international assistance—and some backward moves by the new leadership—were evident in relation to international NGOs (INGOs). Many INGO programs, however, were reviewed, especially those under the Ministry of Home Affairs, now under a new minister who did not regard many of the INGO activities under his purview as being correctly his responsibility, and who generally took a sceptical view of the presence of INGOs. In the ensuing months, some INGO programs suffered as travel restrictions were tightened, and Myanmar officials began to express new doubts about INGO activities that had been previously condoned, if not approved. Little of this was articulated clearly in any policy pronouncements until July 2005, when the government issued draft guidelines for all international programs, prompting the August 2005 withdrawal of the Global Fund for HIV/ AIDS, Tuberculosis and Malaria. The guidelines were criticised in the foreign media and reportedly became the subject of an official complaint by the UN Resident Coordinator, who could have feared a further loss of aid funding. The Global Fund decision was a set-back for international assistance to Myanmar in that it would have only confirmed the Myanmar leadership's cynicism about political bias against Myanmar.[21]

How much these official restrictions will affect international assistance agencies' operations will have to be tested in practice. Earlier, similar restrictions were not always rigorously enforced, but these formal guidelines imply an entrenched disposition to control international assistance, rather than facilitate it. Yet the reality was that, as of mid 2006, only one international humanitarian agency (Médecins sans Frontières) had withdrawn from Myanmar because it could no longer operate effectively. The Humanitarian Dialogue office was also forced to close down, when its head, Leon de Riedmatten, was not able to renew his visa despite playing a prominent advisory role under General Khin Nyunt.[22] While all INGOs were affected by the increased slowness in obtaining government permissions—not helped by the move of government functions to Naypyitaw from late 2005—most were able to continue their basic programs satisfactorily and preferred to hope for improvements in the future. By the middle of 2006, for example, some INGOs had had new memorandums of understanding successfully approved or extended, had expanded their activities and were spending the highest amounts of assistance ever. Others, who chose to maintain a low public profile, claimed not to be experiencing significant disruptions to their activities.[23]

Another serious set-back occurred with the International Committee of the Red Cross (ICRC), whose integrity, impartiality and confidentiality are universally accepted and which the regime had previously found highly useful.[24] By the end of 2005, the ICRC was having unprecedented problems securing access to political prisoners after the Myanmar authorities decreed that they wished to attach their own representatives to the ICRC prison delegates' visits, in breach of long-standing ICRC policy.[25] It seemed that the authorities would have been satisfied if a representative of the para-statal Union Solidarity Development Association (USDA) accompanied ICRC teams, but this was hardly likely to be acceptable. For the first six months of 2006, the ICRC sought unsuccessfully to negotiate a resumption of access to prisoners of security concern on the basis of procedures that were for it universally accepted. As of September 2006, it had not managed to obtain permission to resume its prison visits, but was maintaining its presence in various locations while carrying out its other programs more or less normally.[26]

It was in relation to the ILO, however, that the SPDC's attitudes towards the UN system were at their worst.[27] ILO staff had displayed enormous patience and skill since 2000 and, for a period after 2003, there was some hope for progress as reports of forced labour were transmitted to the ILO Liaison Office for investigation, and some relatively junior officials were punished for ordering forced labour. With the ILO Liaison Office able to report some modest improvements in forced labour in 2003, the ILO's 'engagement' strategy seemed to offer slight prospects for progress. At the annual meeting of the International Labour Conference in June 2004, it was decided not to invoke sanctions, but to renew yet again the ILO's requests that the SPDC respond to its proposals for effective steps to deal with reports of forced labour. In this respect, the ILO showed that enormous persistence and forbearance could produce results for its engagement approach. While the Myanmar government continued to profess its readiness to cooperate with the ILO on ending forced labour, in practice, it repeatedly dragged its feet in implementing effective measures to bring forced labour to an end. The main problem for the government remained the reality that local military forces depended on forced labour to carry out routine administrative and infrastructure works. Moreover, the army was accustomed to exercising its authority over local communities in this way and was not inclined to be dictated to by outsiders on its activities.

Any modest signs of progress on forced labour were reversed in 2005 when the Myanmar authorities announced a policy of prosecuting anyone who made what they considered a 'false complaint' of forced labour, and began to arrest and jail those who reported forced labour. This prevented the ILO Liaison Officer from passing allegations of forced labour that he received to the authorities for investigation, as had been happening for the previous two years. The last straw was when the government in 2005 allowed death threats to be sent to the ILO Liaison Officer and the ILO Facilitator. Making matters worse, in mid 2005, the SPDC also orchestrated (through its para-statal organisations such as the USDA, the War Veterans Association and the Myanmar Women's Affairs Federation) mass meetings across the country attacking the ILO and calling for Myanmar to withdraw from the organisation. Regime spokesmen also openly referred to the possibility of withdrawal from the ILO, in an apparent attempt to challenge the organisation to withdraw from Myanmar and to abandon its attempts to work towards the elimination of forced labour.

Myanmar had never before so actively canvassed the possibility of withdrawing from an international organisation of which it was a member, no matter how serious any disagreements. Official statements at this time were equally an outright contradiction of previous assurances that Myanmar would cooperate fully with the ILO and, if carried out, they would have certainly amounted to the clearest rejection of international norms ever by Myanmar. For its part, the military leadership was certainly aware of the consequences their withdrawal from the ILO would invite, having made their own assessment of the costs and benefits of withdrawal.[28]

Subsequently, the SDPC backed down from this implied threat. This was communicated by the Myanmar Ambassador in Geneva to the ILO,

and mass rallies against the ILO and threats against the ILO Liaison
Officer in Yangon ceased as suddenly as they had begun. Significantly,
on the eve of the June 2006 International Labour Conference, the
Myanmar authorities released two individuals who had been jailed
for reporting forced labour. Hardly surprisingly, the ILO was not to
be easily persuaded that it should let bygones be bygones and, at the
2006 International Labour Conference, it reissued an ultimatum to
the Myanmar government to resume full cooperation with the ILO
by November 2006 or face international sanctions. Having so many
times deferred taking the ultimate decision and given the SPDC the
benefit of the doubt, it seemed that the inevitable 'day of judgment'
for Myanmar had arrived.

The conclusions to be drawn from this long and frustrating hiatus
were that the new SPDC leadership was not ready to cooperate with
the United Nations on political issues; that it fiercely resented the
intrusion of external ideas into Myanmar's affairs; and that it preferred
a self-sacrificing autarchic approach rather than submitting to outside
pressure. Given the pattern of negative developments, it was hardly
surprising that UN Security Council members yielded to pressure from
the United States and the United Kingdom and Myanmar was discussed
informally before Security Council members on 19 December 2005.
Although Myanmar might not constitute a 'threat to international
security' that would warrant a specific Security Council response under
Chapter VII of the UN Charter, it was certainly not cooperating with the
United Nations, and a number of its policies were directly and adversely
affecting neighbouring countries. Bringing Myanmar's situation to the
Security Council had long been a goal of Burmese activists and some
Western countries, but Myanmar had always bitterly opposed this, usually
with the full support of Permanent Security Council Members China
and Russia. Pressure to go to the Security Council intensified after a
report commissioned by former Czech President Václav Havel and South
African Archbishop Desmond Tutu was presented to Secretary-General,
Kofi Annan, in September 2005. Although the Myanmar government
was not alone in criticising this report for its many inaccuracies and

for its lack of objectivity, the absence of any positive developments in Myanmar made it almost inevitable that UN Security Council action of some kind would occur. The issue for the Security Council turned to what specific measures could be endorsed for action.

The only positive note during this period was the May 2006 visit by UN Undersecretary-General for Political Affairs, Ibrahim Gambari, who met not only Head of State, Than Shwe, but Aung San Suu Kyi—the first outsider to meet her since June 2003. Apart from the Secretary-General himself, Gambari was the most senior UN official in several years to take a close personal interest in Myanmar. Gambari was precise and careful in his public comments about his visit. He tried to avoid raising expectations unrealistically and made it clear that his objectives were to improve Myanmar's relationship with the international community and the UN system, and to enable them to better help Myanmar by having 'better access and guarantees'. He said he reached agreement that the United Nations could 'play a role in promoting common ground between the Government and the National League for Democracy (NLD) so that the National Convention could resume in a more inclusive way', and that he saw 'signs of openings' for a commitment by the government 'to re-engage with the international community as partners'.[29] Gambari emphasised the language of conflict resolution and confidence building and, in a subsequent newspaper opinion piece, he called for 'sustained engagement' as the only way to marshall efforts to solve the Myanmar problem.[30]

What induced the SPDC to receive Gambari in May 2006 is not clear, especially as it was Gambari who briefed UN Security Council members in December 2005. While the SPDC might have been seeking to head off further Security Council action by allowing Gambari to meet Aung San Suu Kyi, a single visit could have only limited positive impact. Nevertheless, Gambari's visit was seen as the first sign of a relaxation in the regime's new hard-line approach to political change, but translating it into further concrete progress towards acceptable political change remains a challenge. It is still doubtful that the Myanmar leadership would accept a more pro-active UN role, and probable that it would find support in the Security Council to oppose any UN resolution under Chapter VI of the UN Charter, or any more specific UN mandate in Myanmar.[31] The SPDC has not so far disclosed its intentions, however, now that the Myanmar issue has reached the highest echelons of the United Nations, continued pressure for a more vigorous UN effort is likely.

In September 2006, Myanmar was elevated to the formal UN Security Council agenda, but further action was delayed by the emergence of other more urgent international crises. Although there are expectations of a more effective UN role, the new UN strategy foreshadowed by Gambari had not emerged by October 2006 nor had there been substantial efforts by the United Nations to reopen dialogue with the Myanmar authorities. A uniform policy of sustained engagement by the international community towards Myanmar could hold some hope for a resolution of the long-standing Myanmar problem, but how this position might be reached is far from clear. Unfortunately, in the second half of 2006, the United Nations became preoccupied with more urgent problems, and it is not clear whether the new UN Secretary-General will display the same level of interest in Myanmar as his predecessor.[32]

Prognosis

In 2006, Myanmar was by no means isolated internationally, despite sanctions imposed against it by some and despite its own less-than-complete restrictions on interaction between the Burmese people and the outside world. Much of the interaction between Myanmar and the international community is, however, not designed to achieve—or is intended to actively prevent—change and reform and greater efficiency and openness. These patterns have been exacerbated under the current military leadership, which masks socioeconomic failings behind its exaggerated concerns about sovereignty and security.

With almost no history of clearly successful outside influence being exerted over the highly introverted SPDC, the hopes of the international community to reform Myanmar now rest almost entirely with the United Nations, although China and ASEAN have potentially important roles to play. The United Nations, however, still lacks an overall, specific and detailed strategy and has still not achieved a convincing political consensus in support of a better-defined UN role. Such a role would, of course, need to be backed by funding if it were not to fail. Aid donors need to work together more pro-actively and more transparently than they have in the past to bring a concerted, coherent focus to assistance programs.

Myanmar's international relationships are almost entirely the product of the policies and wishes of the present members of the SPDC and especially Senior General Than Shwe. Yet to change this equilibrium requires first and foremost a change in the attitude of the Myanmar leadership, without which the most carefully designed plans for increased engagement by the international community will fail. No sensible international strategy should, however, be content to be denied the potential benefits for the people of Myanmar from expanded, and better targeted, international engagement. At a time when the United Nations is endeavouring to recalibrate its strategies for assisting Myanmar and averting humanitarian and other crises there, it behooves the international community to get more solidly behind this attempt than it has in the past. Individualistic policies towards Myanmar pursued by great powers or small, countries near or far, will undermine this effort because they will confuse the message that the present Myanmar leadership needs to understand that it has failed the people of Myanmar comprehensively and is no longer entitled to remain at the helm of the country.

A key question remains whether ASEAN can sustain an effective approach to Myanmar that maintains some political integrity and exercises real leverage over the regime. Arguably, apart from China, ASEAN is one of the few sources of effective outside influence over the regime. The current Myanmar leadership is unlikely to permit ASEAN access to Aung San Suu Kyi—as this would enhance the NLD's claims to political legitimacy—but it could undermine its critics by allowing selected representatives of the international community access to all legal opposition groups. At another level of regional economic and social integration, ASEAN can exercise a powerful normative effect on Myanmar (and the other new members of ASEAN), but much greater international assistance is needed for proven ASEAN programs to achieve their full potential in disseminating better governance and pursuing more ambitious outcomes. Myanmar seems to be losing some of the support it once enjoyed inside ASEAN, and the departure of Thai Prime Minister, Thaksin Shinawatra, is also a blow for the Myanmar government, although it might result in Thai policy being more in harmony with those of its ASEAN neighbours. The other key question is whether China will decide that its interests would be served by playing a more substantial political role.

International attitudes to Myanmar will continue to face difficulty in gaining acceptance by the Myanmar government until the international community makes it clear with a single voice that the issues surrounding Myanmar are its legitimate concern. Myanmar's leadership is adept at identifying rifts in international opinion that work to its advantage, and will not stop trying to deflect pressure on it to change policies. Moreover, the capacity of the current Myanmar military leadership to stand stubbornly in the face of international opinion should not be underestimated. This makes it all the more necessary for the international community, led by the United Nations, to draw the Myanmar government into a more focused and managed reconciliation process than has hitherto been the case, with sufficient incentives to persuade the current military leadership to participate fully in such a process. Ultimately, however, success can be achieved only if Myanmar believes that it 'owns' the process, that it has not been imposed from the outside.

Notes

1   The speech was given to National Convention delegates on 22 October 2004 by SPDC Secretary One, Lieutenant-General Thein Sein, who had also been acting as convener of the National Convention. It was reported under the headline 'Change of Prime Minister does not change the government's roadmap agenda' (Thein Sein 2004).

2   This is a grouping of MPs from Cambodia, Indonesia, Malaysia, Singapore and Thailand, set up in November 2004 with support from the Open Society Institute (OSI). See http://aseanmp.org/index.php and the OSI site, http://www.soros.org/initiatives/bpsai/focus_areas/grantee_folder_initiative_view

3   By being unexpectedly vocal about Myanmar, ASEAN undoubtedly elevated the level of discomfort for the Myanmar leadership, but what had it actually achieved? After the 2006 ASEAN Ministerial Meeting, there was still no overall ASEAN strategy of resolving the Myanmar problem or specific ASEAN proposals to encourage Myanmar to move on political reform. ASEAN's inconsistent performance on Myanmar left the perception that it remained a politically weak organisation without the procedures or traditions to deal effectively with political problems. Moreover, the differences in the attitudes of individual ASEAN member countries did not help ASEAN's collective management of the issue. The fact that some ASEAN countries could themselves be criticised on human rights grounds and had internal security provisions not unlike those of Myanmar also reduced ASEAN's credibility and limited the leverage it commanded.

4   Author's conversation with senior Myanmar Foreign Ministry official, March 2005, Yangon.

5   Myanmar's increasing economic dependence on China is described in Kudo 2006.

6   Although it consistently speaks out against any broadening of sanctions against Myanmar, China has not so far been noticeably vigorous in seeking the removal of international financial institutions' sanctions against Myanmar—for example, those imposed by the Asian Development Bank.

7   The impact of the Global Witness report *A Conflict of Interests: the uncertain future of Burma's forests* was magnified because it was also published in Burmese language. It could also have prompted the Myanmar authorities in mid 2006 to start to take an interest in the problem of corruption.

8   Author's conversation with senior Myanmar Foreign Ministry official at the time.

9   For reporting on General Soe Win's visit to China, see McGregor and Kazmin 2006. *The Irrwawaddy Online Edition* article was published in the July 2004 edition. Larry Jagan's article appeared in *Asia Times*, 11 April 2006.

10 Naidu 2004 provides an interesting summary of India's approach to Myanmar.

11 The author's discussions with senior Japanese diplomats in July 2006 confirmed this to be official policy.

12 The Japanese government temporarily suspended new development assistance after the Depayin Massacre in 2003, but resumed it without fanfare in 2004.

13 For example, Japan's trade with Myanmar is well behind that of China, Singapore and Thailand, and Japan is the tenth source of foreign direct investment to Myanmar, whereas for ASEAN as a whole, Japan is second only after the United States. See Japan–ASEAN Centre Investment Statistics at http://www.asean.or.jp/eng.index.html (accessed 8 September 2006).

14 See, for example, Congressional testimony by Assistant Secretary of State for East Asian and Pacific Affairs, Christopher Hill, and Assistant Secretary of State for Democracy, Human Rights and Labor, Barry Lowenkron, on 7 February 2006.

15 *Washington Post* staff writer Glenn Kessler (2005) claimed 'Japan…was especially reluctant to challenge Burma, but Tokyo has abruptly shifted its position'.

16 Japan Center for International Exchange and the University of Helsinki 2006.

17 See various commentaries by Derek Tonkin, especially in his online newsletter, *Burma Perspectives*, 5 July 2005.

18 A notable case is the action by the Netherlands government to prevent the Myanmar Minister for National Planning, U Soe Tha, from attending an ASEM Economic Ministers meeting in the Netherlands in February 2005, even though his visit was permissible under the European Union's Common Position. This led to the cancellation of the meeting, a move calculated to irritate not only the Myanmar regime but its ASEAN colleagues.

19 Verghese Matthews, 2005. Quoted in BurmanetNews, 18 October, no. 2825. Available from editor@burmanet.org.

20 There was, however, no evidence to support some Australian media claims in mid 2006 that Myanmar was seeking nuclear weapons from North Korea.

21 They would recall the long and successful process they conducted to seek funding from the Global Fund for HIV/AIDS, Tuberculosis and Malaria, meeting all criteria through an open process of international bench-marking bids, only to have the substantial $98 million program terminated. For an objective account of this decision, see International Crisis Group 2006. *Myanmar: new threats to humanitarian aid*, Asia Briefing No.58, International Crisis Group, Brussels.

22 de Riedmatten had also carried out the role of the ILO Facilitator, which could have accounted for the Myanmar government's decision not to renew his visa.

23 Communications to the author from Country Program Managers of one large and one small INGO with continuing Myanmar programs in July 2006.

24 But it should not be forgotten that once before, in 1995, the ICRC had withdrawn from Myanmar in protest against unacceptable restrictions placed on its activities.

25 In September 2006, the ICRC spokesperson in Yangon for the first time expressed publicly concern that ICRC prison visits had still not been resumed, repeating this in December and February. The ICRC does not normally publicise its problems.

26 Confidential communication to author, July 2006.

27 I am indebted to ILO Liaison Officer Richard Horsey for help with the factual accounts in these paragraphs, although the judgments here are entirely the author's.

28 In the early 2000s, the Myanmar government constituted its own advisory team on the forced labour issue made up mainly of retired Myanmar diplomats. Author's conversations with members of this team.

29 Press Conference on Myanmar, 25 May 2006, Department of Public Information, United Nations, New York.

30 See Gambari's own account (2006).

31 The Myanmar leadership might be concerned about China's position after China decided to support UN Security Council action against North Korea over its nuclear test in October 2006.

32 Gambari's own term will conclude early in 2007, and much will depend on the activism of his successor on Myanmar.

References

Albar, S.H., 2006. 'It is not possible to defend Myanmar', *Asian Wall Street Journal*, 24 July.

Asian Survey. Annual surveys of developments in Myanmar, University of California, Berkeley.

Bert, W., 2004. 'Burma, China and the USA', *Pacific Affairs*, 77(2) (Summer):263–82.

DLA Piper Rudnick Gray Cary, 2005. *Threat to the Peace Report*, Report commissioned by Václav Havel and Desmond Tutu, DLA Piper Rudnick Gray Cary, New York. Available from: http://www.unscburma.org/Docs/Threat%20to%20the%20Peace.pdf.

Gambari, I., 2006. 'A crack in the Burmese door', *International Herald Tribune*, 21 June.

Ganesan, N., 2005. 'Myanmar's foreign relations: reaching out to the world', in K.Y. Hlaing, R. Taylor and T.M.M. Than (eds), *Myanmar: beyond politics to societal imperatives*, Institute of Southeast Asian Studies, Singapore.

Global Witness. 2003.'A conflict of interests: the uncertain future of Burma's forests', report by Global Witness. Available from: http://www.globalwitness.org/reports/index.php?section=burma (accessed 24 April 2006).

Haacke, J., 2005. '"Enhanced interaction" with Myanmar and the project of a security community: is ASEAN refining or breaking with its diplomatic and security culture?', *Contemporary Southeast Asia*, 27(2):188–210.

——, 2006.'Myanmar's foreign policy', Adelphi Paper No.381, Institute of International and Strategic Studies, London.

International Crisis Group, 2006. *Myanmar: new threats to humanitarian aid*, Asia Briefing No.58, International Crisis Group, Brussels.

James, H., 2004. 'Myanmar's international relations strategy: the search for security', *Contemporary Southeast Asia*, 26(3):530–53.

Jagan, L., 2006.'Myanmar woos China, Russia', *Asia Times Online*, 12 April. Available from: http://atimes01.atimes.com/atimes/Southeast_Asia/HD12Ae04.html

Japan Center for International Exchange and the University of Helsinki, 2006. *ASEM in its Tenth Year—an evaluation of ASEM in its first decade and an exploration of its future possibilities*, Japan Center for International Exchange and the University of Helsinki, Tokyo and Helsinki.

Kessler, G., 2005. 'US sees Burma as "test case" in Southeast Asia', *Washington Post*, 28 December 2005.

Kudo, T., 2006. *Myanmar's economic relations with China: can China support the Myanmar economy?*, Discussion Paper No. 66 (July), Institute for Developing Economies, Tokyo.

Kurlantzick, J., 2002. 'Can Burma reform?', *Foreign Affairs*, November/ December.

Matthews, V., 2005. 'Myanmar makes Europe a reluctant player', *New Straits Times*, 18 October.

McGregor, R. and Kazmin, A., 2006. 'Burma's stability a concern for China', *Financial Times*, 14 February.

Naidu, G.V.C., 2004a. 'Whither the look East policy: India and Southeast Asia', *Strategic Analysis*, 28(2), April–June:331–346.

——, 2004b.'Looking East: India and Southeast Asia', paper presented at the Institute for International Relations/Institute for Defence Studies and Analyses Second Roundtable, 27–28 October, Taiwan.

Pereira, R., 2005. 'The Fifth Asia–Europe Meeting summit: an assessment', *Asia Europe Journal*, No.3:17–23.

Takeda, I., 2001. 'Japan's Myanmar policy: four principles', *Gaiko Forum* (in English), Summer.

Thein Sein, 2004. 'Change of Prime Minister does not change the government's roadmap agenda', *New Light of Myanmar*, 23 October.

Transnational Institute, 2006. *Drug Policy Brief*, No.17 (May), Burma Centrum Nederland:10–13.

6 Myanmar's economy in 2006

Sean Turnell

Myanmar was likely to experience moderate but superficial economic growth through 2006. The country's ruling military regime, the self-styled State Peace and Development Council (SPDC), has claimed GDP growth rates in excess of 10 per cent per annum for almost a decade. If true, this would make Myanmar one of the world's fastest and most consistently growing economies. These claims are without foundation, but a growth rate of between 1.5 and 4 per cent was not beyond reach for 2006. Such growth, however, would primarily be a consequence of the high prices Myanmar can now command for its exports of natural gas, and from greater export volumes of gas from new fields currently being brought on stream.

In every other respect, Myanmar's economy will continue to under-perform, in terms of its own potential and relative to that of its neighbours and peers. Indeed, to a large degree, Myanmar's probable rising gas exports will bring about unfortunate consequences for the country—allowing the SPDC to postpone the economic and political reforms the country needs if its people are to enjoy any measure of economic security. Myanmar, in short, will likely experience a 'gas curse' every bit as inimical to good economic policymaking as has often been a by-product of oil elsewhere.

Myanmar is one of the poorest countries in Southeast Asia, yet, only 50 years ago, it was one of the wealthiest. The dramatic turn around of Myanmar's fortunes is the product of a state apparatus that for decades has claimed the largest portion of the country's output, while simultaneously dismantling, blocking and undermining basic market institutions. The excessive hand of the State—which for many years was wedded to a peculiar form of socialism—has manifested itself in a number of maladies that are the direct cause of Myanmar's current poverty.

- Myanmar's military regime has, in the 40 years it has been in power, systematically dismantled the fundamental economic institutions—effective property rights, contract enforcement, the measures that define the 'rules of the game' for efficient economic transactions—that history tells us are necessary for sustainable long-term growth.

- Macroeconomic policymaking in Myanmar is arbitrary, often contradictory and ill-informed.

- The government's claim on Myanmar's real resources greatly exceeds its ability to raise revenue through taxation. As a consequence, like many such regimes around the world and throughout history, the SPDC resorts to the printing press to 'finance' its own expenditure. Inflation and monetary chaos have been the predictable consequences.

- Myanmar has a currency, and a financial system, that is widely distrusted. People in Myanmar store their wealth in devices designed as a hedge against inflation and uncertainty. As a result, financial intermediation is underdeveloped and the allocation of capital is distorted. In 2006, Myanmar was still recovering from a major banking crisis that took place in 2002–03.

- Rent seeking through state apparatus offers the surest route to prosperity in Myanmar, at the expense of enterprise. Myanmar's leading corporations are mostly owned and operated by individuals 'connected' to the government, and often serving and retired military officers. Corruption is endemic.

- Important sectors of Myanmar's economy are starved of resources. Negligible spending on education and health have eroded human capital formation, and reduced economic opportunities. Agriculture, which provides the livelihood for the majority of the people of Myanmar, is chronically starved of critical inputs.

- The military regime's economic mismanagement means that Myanmar attracts little in the way of foreign investment. What does arrive is concentrated in the gas and oil sectors, and other extractive industries. Little employment is generated from such investments, and there is little in the way of technology or skill transfer.

Such then are some of the broad factors that inform Myanmar's current economic circumstances. This chapter details more closely specific sectors of Myanmar's economy, their current condition and immediate prospects.

Economic growth

In February 2006, Myanmar's Minister of National Planning and Economic Development, Soe Tha, announced that his country's growth rate for 2005 would be 12.2 per cent.[1] This topped even 2004's stellar growth of 12 per cent and made Myanmar (certain small oil-producing countries excepted) the fastest growing economy in the world.

Table 6.1   Claimed annual GDP growth rates, 1999–2005 (per cent per annum)

	1999	2000	2001	2002	2003	2004	2005
GDP growth	10.9	13.7	11.3	10.0	10.6	12.0	12.2

Sources: Asian Development Bank, 2004. *Asian Development Outlook 2004*, Asian Development Bank, Manila; Asian Development Bank, 2005. *Asian Development Outlook 2005*, Asian Development Bank, Manila.

Stating anything definitive with respect to economic growth in Myanmar is fraught with the difficulties characteristic of a country in which the official statistics are notoriously unreliable, and where collecting routine data otherwise is difficult. Myanmar does not publish national accounts statistics and the only growth data that are made available are those that accompany ministerial statements such as the one above. Nevertheless, we can be sure that economic growth in Myanmar is well below the minister's claims. His boast is greatly at odds with even the most cursory glance at the economic circumstances on the ground in Myanmar, circumstances that point to ever deeper levels of poverty for the average citizen, and to an economy that at worst is on the verge of collapse, and at best cycles through bare subsistence.

Table 6.2   Economic growth estimates, 2001–2006 (per cent per annum)

	2001	2002	2003	2004	2005	2006
Growth estimates	5.3	5.3	–2.0	–2.7	3.7	1.8

Source: Economist Intelligence Unit, 2006. *Burma (Myanmar): country report*, May, Economist Intelligence Unit, London:5.

More substantially, however, we can dispute the minister's claims through various proxy measures and indicators of economic growth. For instance, if Myanmar were truly growing along the lines claimed by the SPDC, one would expect to see it using more productive resources: energy, land, labour, capital and so on. We do not see this. Indeed, as the Asian Development Bank (ADB 2005:30) notes, electricity usage in Myanmar fell by 32.4 per cent in 2004–05. Among other indicators, in the same period, cement output fell 8.5 per cent, sugar production fell by 2 per cent and credit extended to the private sector (Table 6.3 below) was recovering only fitfully from its collapse the year before (and accordingly was lower than in years of slower claimed growth). In 2005, it was likely that manufacturing as a whole—the sector contributed just more than 10 per cent of GDP—contracted, not a result one would expect to see for an economy growing in double digits (Economist Intelligence Unit 2006:18).

In addition to these 'internal' proxies, however, if Myanmar were growing at the rates claimed by the SPDC, we would also presume to see certain patterns in its economy that history tells us to expect of rapidly growing economies (Bradford 2004). We should see less reliance on agriculture, greater reliance on industry and even the emergence of services. Of course, these are long-term patterns, but shorter-term trends are generally at least consistent with them in countries that truly have enjoyed high growth (and for which the Asian 'tiger' economies and China are exemplary). Myanmar displays none of these structural dynamics. Indeed, as demonstrated by Bradford (2004), agriculture has assumed a greater role in Myanmar's economy in recent years. In short, either Myanmar's claimed economic growth numbers are greatly at odds with reality, or the country has truly found a unique path to economic prosperity.

An alternative set of growth numbers (Table 6.2), more consistent with my critique here (and with Myanmar's recent economic history), has been estimated by the Economist Intelligence Unit (Economist Intelligence Unit 2006:5).

As can be seen from the growth estimates, moderate economic growth returned to Myanmar in 2005 and this was likely to continue through 2006. Such growth is driven by the increasing global demand for energy that has pushed up the price of natural gas. Myanmar currently exports natural gas only to Thailand in sizeable quantities, but new projects are being brought on stream via a series of deals with Chinese, Indian and South Korean investors. Increasing gas prices and export volumes caused Myanmar's trade balance to turn positive in 2005 (Economist Intelliegence Unit estimate: 4.4 per cent of GDP), and it was this contribution that was responsible for the country's estimated positive rate of economic growth overall. Contributions from agriculture remain flat (despite relatively good harvests), while other sectors of the economy—manufacturing, transport, services and tourism—are likely to detract from economic growth. These sectors faced particular downside risks through 2006, ranging from high oil prices, potential avian influenza outbreaks and political unrest at home and abroad (especially Thailand) to capricious policy changes, consumer boycotts and the possibility of increased economic sanctions.

Macroeconomic policy

Fiscal policy

Macroeconomic policymaking in Myanmar is coloured by one overwhelming fact: the irresistible demands of the State on the country's real output. These demands far exceed the State's ability to raise taxation revenue and, accordingly, have led to a situation in which the State 'finances' its spending by the simple expedient of selling its bonds to the central bank. This policy (in economics parlance, 'printing money') distorts every other aspect of policymaking in Myanmar. Fiscal policy is concerned simply with the raising and spending of funds, monetary policy likewise with keeping interest rates sufficiently low (as will be examined, negative in real terms) to minimise financing costs. Neither plays a counter-cyclical or developmental role.

The demands of the State on Myanmar's financial resources swamp all others (Table 6.3). Central bank lending to the government is the favoured device for financing government expenditure. Yet, as can also be seen from the data above, the State is a borrower from Myanmar's commercial banks. The latter provide the private sector with little more than one-quarter of the funds that Myanmar's financial system provides to the central government. The small amount of government bonds held by the general public, an infinitesimal proportion—substantially less than 1 per cent—of the bonds sold to the central bank, is indicative of the confidence they hold in such state-created financial assets.

In recent years, the SPDC has introduced dramatic increases in the taxes it levies. This was especially the case with respect to customs duty revenues, which rose by more than 500 per cent in 2004–05, and on current trends would increase further in 2006. The rise in customs duty revenues came via a mix of factors—including increases in duty rates

Table 6.3   State share of Myanmar's financial resources, selected indicators, 1999–2006 (kyat millions)

	Central bank lending to government	Commercial bank lending to government	Commercial bank lending to private sector	Public holdings of government bonds
1999	331,425	12,460	188,149	378
2000	447,581	36,159	266,466	463
2001	675,040	40,985	416,176	504
2002	892,581	43,248	608,401	563
2003	1,262,588	35,546	341,547	544
2004	1,686,341	89,217	428,391	505
2005	2,165,154	100,358	570,924	**446
2006*	2,281,046	125,983	563,769	n.a.

Notes: * as at end February ** as at end June (2006 data unavailable) n.a. not applicable
Sources: International Monetary Fund (IMF), 2006. *International Financial Statistics*, various issues, International Monetary Fund, Washington, DC. Myanmar Central Statistical Office (MCSO), 2006. *Selected Monthly Indicators*, Myanmar Central Statistical Office, Rangoon. Available from http://www.csostat.gov.mm

Table 6.4   Customs duty revenues, 2002–2006 (kyat millions)

	Duty revenues ('normal' trade)	Duty revenues ('border' trade)	Total customs revenue trade
2002	5,826.9	29.9	5,856.8
2003	4,554.3	136.9	4,691.2
2004	3,941.0	90.1	4,031.1
2005	11,822.5	9,030.5	20,853.0
2006*	3,941.2	636.6	4,577.8

Note: * April to June
Source: Myanmar Central Statistical Office (MCSO), 2006. *Selected Monthly Indicators*, Myanmar Central Statistical Office, Rangoon. Available from http://www.csostat.gov.mm/

and relevant exchange rate formulae (more on which below), as well as a crack-down on corruption (real and imagined) within the Customs Department.[2] The effect on duties raised from so-called 'cross-border' trade (mostly with China and Thailand) was particularly dramatic. Notwithstanding the phenomenal increases in customs duty revenues, however, total central government tax revenue in the fiscal year 2004–05 (of K278,024 million) continued to fall well short of government expenditure (Economist Intelligence Unit 2006:17). The SPDC does not publish data on its spending, but given that new advances to the government from the central bank came to K378,697 million in roughly the same period, it is reasonable to assume that taxes account for little more than 40 per cent of government spending.

Finally, the sudden decision by Myanmar's government in April 2006 to increase the salaries of civil servants dramatically (more on which below) will only exacerbate the country's chronic fiscal imbalances. The decision seems to have been made with little concern for how these pay rises might be paid for. A similar series of pay increases, likewise made with little consideration for Myanmar's fiscal position, were granted in April 2000.[3]

Monetary policy

Monetary policy in Myanmar is formally the responsibility of the Central Bank of Myanmar (CBM). A number of factors, however, determine that it is incapable of exercising effective influence over monetary conditions in Myanmar. The first and most simple of these is that Myanmar has in place interest-rate controls that cap lending rates at 18 per cent per annum, and do not allow deposit rates to fall below 9 per cent per annum. These rates, and the rate at which the CBM will provide funds to the commercial banks (the so-called 'Central Bank Rate', currently at 12 per cent), had not changed for a number of years until they were suddenly increased on 1 April 2006 (more on which below). Given that Myanmar's inflation rate was (conservatively) put at just more than 20 per cent in 2005, this implies that 'real' interest rates in Myanmar remain substantially negative (Economist Intelligence

Unit 2006:5). The motivation for locking in such rates (which result in substantial distortions in capital allocation) is to minimise the interest rates to be paid on government debt. Currently, three and five-year Burmese government bonds have fixed yields of 8.5 and 9 per cent respectively (MCSO 2006). The distrust of Myanmar's currency, the kyat, has created parallel foreign currency spheres in Myanmar, and these are also beyond the influence of the CBM. Finally, it perhaps goes without saying that the CBM does not enjoy operational independence from the State.

As noted above, in April 2006, the CBM suddenly announced that it would increase the Central Bank Rate to 12 per cent per annum, and in so doing allow the commercial banks to charge up to 18 per cent on loans, and pay no less than 9 per cent on deposits (in practice, most charge 17 per cent on loans, and pay 12 per cent on deposits). The CBM does not make statements on monetary conditions in Myanmar, but the timing of the move is revealing, coinciding precisely with the equally sudden increase in civil servants' salaries noted earlier. Simply—and although Myanmar's primitive financial system makes irrelevant the standard 'tool-box' of central bank monetary policy devices (open market operations, rediscount facilities, repurchase agreements and so on)—the increase in the Central Bank Rate does seem to have been in order to 'signal' that the government did not want to see an acceleration in inflation. Of course, since the move is devoid of substance, it will have little impact.

In the absence of standard monetary policy instruments, Myanmar's monetary authorities (the CBM in reality is subservient to the Ministry of Finance and Revenue, as well as to senior members of the SPDC) have resorted once more to less orthodox measures in the attempt to control inflationary pressures in the economy. These include rationing (of gasoline and critical foodstuffs), increased government subsidies on certain commodities critical in household expenditure, and arbitrary fines on traders deemed to be profiteering. Equally representative of Myanmar's inflation strategy are the issuing of orders and exhortations. In May 2006, for instance, Minister of National Planning and Economic

Development, Soe Tha, declared that Myanmar's 'inflation rate should not be allowed to increase into double digits and we should make an effort to see to it that inflation is no more than 5 per cent'.[4] Such a declaration, commonplace in the era of the 'command economy' in Myanmar (from 1962 up to about 1988), sits rather oddly with the SPDC's erstwhile objective of creating a market economy.

Exchange rate

Myanmar has a fixed exchange rate policy that officially links the kyat to the US dollar at a rate of approximately K6:US$1.[5] This official rate, however, is just one of a number of exchange rates applicable to Myanmar's currency. The most important of these rates, and the only one relevant to the people 'on the street' in Myanmar, is the black market or unofficial rate. In September 2006, this rate stood at about K1,350: US$1, more than 200 times below the official standard. This rate is, of course, subject to daily, even hourly, fluctuation according to the perceptions of informal currency dealers regarding Myanmar's prospects. Wild swings in the unofficial rate are reasonably frequent, to which the SPDC's counter is invariably to order the rounding up of a cohort of foreign exchange dealers. As a consequence of United States sanctions imposed on Myanmar, the SPDC has employed various coercive measures to try to discourage the use of the US dollar, in favour of the euro, the Singapore dollar, the Thai baht and the Japanese yen. These measures have had limited success, and the US dollar remains a highly prized store of value (especially, in this context, 'new' US$100 bills).[6]

Table 6.5   Indicative (unofficial) exchange rates, 1997–2006 (kyat/US$1)

	1997	1998	1999	2000	2001	2002	2003	2004	2005	2006
Exchange rates	240	340	350	500	650	960	900	1,000	1,300	1,240*

Notes: *as of November (estimates based on information supplied to the author by bankers in Yangon)
Source: Author's calculations

In addition to its sometimes wild fluctuations, the unofficial value of the kyat has been in decline for some time, and in this sense it acts as something of a barometer of the state of Myanmar's macroeconomy. Table 6.5 records its declining value *vis-à-vis* the US dollar in the past decade.

In addition to the official and unofficial exchange rates, there are other, semi-official rates that apply depending on the counterparties and circumstances. For instance, a rate of K450:US$1 applies formally for all funds brought into Myanmar by UN agencies and international non-governmental organisations (INGOs).

Like many other economic decrees in Myanmar, however, this one is honoured primarily in the breach, and the affected institutions have devised a number of innovative schemes to get around the formal rule, which otherwise penalises them and reduces their available resources. Until June 2006, this exchange rate also applied for the purposes of excise calculation on imports into Myanmar. In June, however, this 'dutiable' exchange rate was suddenly increased to K850/US$1, nearly doubling Myanmar's effective import tax even though the nominal tax rate remained unchanged (at 25 per cent on most items) (Lwin 2006).

Myanmar's multiple and divergent exchange rates are the public face of the country's macroeconomic malaise. They also provide extraordinary opportunities for rent seeking and opportunistic currency deals. It is clear, for instance, that having access to foreign currency at anything close to the official exchange rate presents the recipient with the potential of immediate windfall gains. Reforming and unifying Myanmar's exchange rate regimes, which should mean allowing the kyat to 'float', should be a first-order priority in any future reform program. Such reforms have been advocated regularly by the International Monetary Fund (IMF) in its 'Article IV' consultations with Myanmar, seemingly to no avail.[7]

'Capricious' policymaking

One of the most damaging features of macroeconomic policymaking in Myanmar (of all types), is that it is often made in ways that, to observers and those directly affected, appears highly capricious, arbitrary, selective and even simply irrational. Examples of such decision making are legion, and the following are but a small but indicative recent sample.

- Effective from 1 April 2006, Myanmar's Ministry of Finance and Revenue suddenly announced salary increases for the nation's civil servants and military personnel of between 500 and 1,200 per cent. The announcement did not say how the pay increases were to be funded. In expectation that inflation was likely to accelerate as a consequence of these pay rises, traders in Yangon and elsewhere pre-empted matters and began lifting prices as soon as the announcement was made. There was also a flight from the kyat and hefty increases in the price of gold, foreign currencies and other traditional inflation hedges.

- In October 2005, the SPDC suddenly announced an eightfold increase in the retail price of gasoline.

- Various announcements were made throughout 2005 that exporters/importers in Myanmar were to henceforth use the euro rather than the US dollar in their transactions.

- The (numerous) changes to tax and duty levies on commodities included, in 2004, not only dramatic hikes in import duties on certain (mostly consumer) goods, but changes in the exchange rates applicable for their calculation. As noted earlier, in June 2006, the import duty exchange rate once more suddenly increased—to K850/US$1.

- There have been reflexive cycles of relaxation and restriction on border trade, sometimes in connection with 'purges' of corrupt officials.

- A sudden announcement was made in 2005 that Myanmar's administrative capital would relocate from Yangon to Pyinmana (Naypyitaw). There is little to suggest that the economic dislocation costs of the move (to the government itself, and those who must deal with it) were seriously considered.

External sector

Trade

It is only from the external sector that any growth in Myanmar's economy is apparent—or likely. Driven by rising gas export prices and volumes, Myanmar recorded a trade surplus in 2004 of more than

Table 6.6   Myanmar's external sector, selected indicators, 1999–2005
(US$ million)

	Goods exported	Goods imported	Current account balance
1999	1,293.9	2,181.3	−284.7
2000	1,661.6	2,165.4	−211.7
2001	2,521.8	2,443.7	−153.5
2002	2,421.1	2,022.1	96.6
2003	2,709.7	1,911.6	−19.3
2004	2,926.6	1,998.7	111.5
2005*	836.6	364.5	296.6

Note: * as at end of first quarter
Source: International Monetary Fund, 2006. *International Financial Statistics*, various issues, International Monetary Fund, Washington, DC.

Table 6.7   Composition of exports, 2002–2005 (kyat million)

Export type	2002	2003	2004	2005 (as of end of April)
Gas	4,247	5,919	3,334[8]	3,461
Teak and other woods	1,880	1,874	2,149	810
Pulses	1,898	1,744	1,407	503
Garments and textiles	2,985	2,973	1,298	368
Shrimp and fish products	829	829	1,003	230
Metal and ore	288	288	503	220
Rice	754	633	112	90
Rubber	76	89	81	61

Sources: Economist Intelligence Unit, 2004. *Burma (Myanmar): country profile*, Economist Intelligence Unit, London. Economist Intelligence Unit, 2005. *Burma (Myanmar): country profile*, Economist Intelligence Unit, London. Economist Intelligence Unit, 2006. *Burma (Myanmar): country report*, May, Economist Intelligence Unit, London. Myanmar Central Statistical Office, 2006. *Selected Monthly Indicators*, Myanmar Central Statistical Office, Rangoon. Available from http://www.csostat.gov.mm

US$900 million. For the first three months of 2005—the latest data publicly available—the surplus in this item stood at nearly US$470 million (IMF 2006). With gas prices rising in 2005 and greater volumes likely to have been shipped, a large trade surplus slightly in excess of US$1 billion for the year as a whole was expected. For 2006, this trend was likely to continue, with the Economist Intelligence Unit (EIU) (2006:5) predicting an annual trade surplus of US$1.2 billion. It will be noted from Table 6.8 below, however, that imports into Myanmar have been falling in recent years. This seems unlikely to continue for much longer, especially as Myanmar's imports required infrastructure to develop the new gas fields that have been the subject of recent deals (Table 6.6). To a considerable extent, Myanmar's trade surpluses are offset by deficits in services and in income payments—all of which diminish the overall surplus on current account. This trend likewise will continue into the future—driven by the repatriation of profits by the (largely foreign) firms investing in Myanmar's energy sector.

Gas exports exceeded that of the whole of 2004 by the end of the first quarter of 2005. So far, most of this gas is sourced from the existing Yadana and Yetagun fields (almost all of which is exported to Thailand), but this will shortly be joined by gas piped from sites soon to come on stream, including that of the (offshore) Korean/Indian/ Burmese ventures in Rakhine State (more on which below). The vast bulk of Myanmar's exports are from extractive industries of various types (Table 6.7).

Worryingly, as the EIU (2006:24) notes, exports of Burmese teak are likely to be substantially understated when one considers the pervasiveness of illegal logging in the country. Myanmar's exports of garments and textiles have contracted substantially in the past two years, a function of economic sanctions, consumer boycotts and, not least, by the ending of the Multi-Fibre Agreement that saw China increase its share of the global garment industry at the expense of countries such as Myanmar (Turnell 2006).

Foreign investment

Myanmar is not a large recipient of foreign direct investment (FDI). The country is regarded as a highly risky destination for foreign investment and a difficult location in which to do business. In a recent report on economic freedom, the Washington-based Heritage Foundation ranked Myanmar third from the bottom (in front of only Iran and North Korea) with regard to restrictions on business activity. According to the foundation, 'pervasive corruption, non-existent rule of law, arbitrary policy-making, and tight restrictions on imports and exports all make Myanmar an unattractive investment destination' (Miles et al. 2006:125).

Recent FDI in Myanmar was directed overwhelmingly to the gas and oil sectors (Table 6.8). Very little FDI made its way to industry, and even less to agriculture (which received FDI of a mere US$34.4 million

Table 6.8   Foreign direct investment flows, sector and source, 2003–2005 (US$ million)

	2003	2004	2005 (as of end April)
Sector			
Gas and oil	44.0	54.3	142.6
Real estate	-	-	31.3
Mining	3.4	1.5	6.0
Manufacturing	13.2	2.8	3.5
Transport	-	30.0	-
Agriculture and fisheries	26.4	2.6	-
Source country			
China (including Hong Kong)	12.9	2.8	126.6
Thailand	-	22.0	29.0
Japan	-	-	2.7
Malaysia	62.2	-	-
South Korea	0.3	34.9	-
United Kingdom	-	27.0	-

Sources: Economist Intelligence Unit, 2004. *Burma (Myanmar): country profile*, Economist Intelligence Unit, London; Economist Intelligence Unit, 2005. *Burma (Myanmar): country profile*, Economist Intelligence Unit, London; Economist Intelligence Unit, 2006. *Burma (Myanmar): country report*, May, Economist Intelligence Unit, London.

since the 'opening' of Myanmar 17 years ago).[9] In terms of source countries, the traditional largest investors in Myanmar—Singapore and Thailand—have in recent times been overshadowed by China. This trend is likely to continue, albeit with China joined by greater investment in Myanmar's gas sector by Indian and Korean investors.

New gas ventures

In the past few years, several significant Myanmar–foreign joint ventures concerned with exploring and exploiting Myanmar's large reserves of natural gas have been announced. Thus far, most of the funds spent by foreign investors in this context have been for exploration and preliminary drilling. With the results of this exploration proving highly positive, however, and in the face of growing demand and rising prices for natural gas, these ventures are now moving towards active exploitation and production.

The most lucrative of Myanmar's new gas fields are the so-called '*Shwe*' ('gold' in Burmese) and '*Shwephyu*' ('white-gold') fields, offshore from Sittwe, in Rakhine (Arakan) State. In the terminology applied by Myanmar's Ministry of Energy, these fields (which are located in Myanmar's far northwest, adjacent to the Bay of Bengal) are referred to collectively as the 'A1-Block'. They are currently being explored by a consortium that comprises Myanmar's state-owned Myanmar Oil and Gas Enterprise (MOGE), South Korea's Daewoo International Corporation (which owns 60 per cent of the foreign component of the venture), the Korean Gas Corporation (10 per cent), the Gas Authority of India Limited (GAIL)—majority owned by the Indian government, which owns 10 per cent, and the Oil and Natural Gas Corporation (ONGC) Videsh, also from India, which holds the remaining 20 per cent. Under the joint venture, MOGE is entitled to 50 per cent of the gas extracted. In early 2004, the consortium reported that it believed the field had between 14 and 20 trillion cubic feet of recoverable gas reserves.[10] For some time, however, the A1 venture has been embroiled in a controversy about the ultimate destination of the subsequent gas exports. India was long the presumed customer, but a problem emerged in that the shortest route for a pipeline to the country would have to pass through Bangladesh, which demanded certain trade and other concessions before construction could proceed. Negotiations between the two countries proceeded for a time, before apparently breaking down irrevocably. In late December 2005, the Indian government announced that any proposed pipeline would now bypass Bangladesh, coming onshore at Sittwe and passing through Myanmar's Chin State, before terminating in Kolkata—a route some 250 miles longer than that via Bangladesh (EIU 2006:22–3).[11]

Throughout the protracted pipeline negotiations, the Myanmar government warned that other potential customers for the output of the A1-Block were ready to step in. So it proved, and in February 2006 it came to light that a memorandum of understanding had been signed (some time in November 2005) between PetroChina and MOGE to sell China 6.5 trillion cubic feet of gas from MOGE's share of the block (Fullbrook 2006). This move prompted a response from the Korean and Indian joint-venture partners to ensure their share of gas delivery. In January 2006, Daewoo International announced it had secured 3.6 trillion cubic feet of gas from the A1-Block, and would invest US$120 million in developing the necessary infrastructure.[12] In March 2006, during a visit to Myanmar by India's President, A.P.J. Abdul Kalam, a memorandum of understanding was signed between the two countries under terms for final gas exports identical to those reached with China (*The Economic Times* 2006). In August 2006, in what will by no means be the last word from prospective suitors for the output of the Shwe fields, Thailand's state-owned PTT Exploration and Production oil and gas firm entered the fray, seeking a 20-year supply deal from the A1-Block to complement its Yetagun and Yadana arrangements. The first gas exports from the A1-Block are due to flow from 2009.

In addition to the A1-Block, other gas fields off Rakhine State are being explored. The most significant of these, the so-called 'A3-Block', are being surveyed by MOGE and three of the four foreign joint-venture partners involved in the A1-Block (Daewoo, GAIL and ONGC Videsh, with stakes of 70, 10 and 20 per cent respectively). The A3-Block is estimated to have gas reserves of three trillion cubic feet (*Xinhua News* 2006; *Oil and Gas Journal Online*, 2006a).

Together, the A1 and A3 Blocks off Rakhine State are significantly larger than the Yadana and Yetagun fields, which currently provide the bulk of Myanmar's gas exports. The latter fields, located in the Andaman Sea, have proven gas reserves of 6.52 trillion cubic feet and 3.2 trillion cubic feet respectively (*Oil and Gas Journal Online* 2006b). In other words, they are approximately half the size of the fields discovered off Rakhine. Yadana was a joint venture between MOGE and Total Oil (France), Unocal (United States) and PTT Exploration and Production (Thailand). The Yetagun fields were developed by MOGE, Premier Oil (United Kingdom, now replaced by Petronas of Malaysia) and Nippon Oil (Japan).

Foreign exchange reserves

Myanmar's trade surpluses and (to a lesser degree) the flows of FDI have swelled the country's official foreign exchange reserves—from US$265 million in 1999 to more than US$770 million today (Table 6.9). The latter number, however, is still very low by global or even regional standards. Table 6.9 contains a sample of countries with which, for a

Table 6.9   Foreign exchange reserves, selected countries, 2000–2005 (US$ million)

	Myanmar	Thailand	Cambodia	South Korea	Vietnam
2000	223	32.016	502	96,131	3,417
2001	400	32,355	587	102,753	3,675
2002	470	38,046	776	121,345	4,121
2003	550	41,077	815	155,284	6,224
2004	672	48,664	943	198,997	7,042
2005	770	50,728	939	210,317	8,602

Source: International Monetary Fund, 2006. *International Financial Statistics*, various issues, International Monetary Fund, Washington, DC.

variety of reasons, Myanmar might be compared. It can be seen that Myanmar has, by some margin, the lowest level of reserves 'comfort', even when compared with tiny and poor Cambodia. Of course, Myanmar's foreign assets must also be set against its foreign liabilities. These currently stand at about US$7 billion (or about 10 times the size of the country's reserves), and consist for the most part of defaulted loans to the World Bank and other multilateral lenders (IMF 2006).

Monetary and financial sector

Myanmar's financial system—a mix of state-owned institutions, 15 surviving privately owned banks in varying degrees of health and a dominant informal sector—is failing to meet the country's need for capital. As noted in Table 6.3, the largest claimant on credit creation in Myanmar is the State. Private-sector trade and industry in Myanmar can access some credit from the private banks, but the macroeconomic instability of the country means that much of this is of a short-term nature only, and is concentrated in such inflation-hedging sectors as real estate and precious metal and stone trading. Long-term credit for industrial development is almost completely non-existent. Personal credit in Myanmar is available from formal financial institutions for a handful of well-connected élites, but for the average person in Myanmar credit is supplied by friends, relatives or, less agreeably, the local money-lender—for time immemorial, a ubiquitous presence in the country (Turnell 2006). For agriculturalists in Myanmar, the availability of credit is especially dire. According to a recent Food and Agriculture Organization (FAO) survey (2004:141), 80 per cent of Myanmar's agriculturalists were without access to formal credit of any kind.

As recently as 2002, however, it was possible to entertain some optimism with regard to the financial system in Myanmar, particularly with respect to the private banks. These had emerged only since 1990 and the implementation of certain financial-sector reforms (principally the Financial Institutions of Myanmar Law and the Central Bank of Myanmar Law, both promulgated in 1990). By 2002, the private banks appeared to be growing strongly and, among the largest of them, the creation of a degree of trust and even 'brand recognition' seemed apparent. Beneath the surface, however, all was not well. Myanmar's interest rate restrictions (noted above) greatly hampered the private banks in traditional intermediation (taking in deposits and making loans), forcing them into activities of high risk and questionable legitimacy. That said, some of the private banks were established in the first instance precisely to conduct and disguise unorthodox and criminal activity (regarding the latter, the laundering of narcotics money especially), while others were little more than corporate cash boxes for various entities connected with the regime. In 2002, all of this bubbled to the surface as a financial crisis engulfed Myanmar.

At the centre of Myanmar's 2002–03 financial crisis was a banking collapse that was almost archetypal of such phenomena. Beginning in November 2002, long lines of anxious depositors formed outside the banks, a spectacle that rapidly swelled into a classic bank run. From this moment on, the response of the relevant monetary authorities in Myanmar (principally the CBM) was almost wholly destructive. Late

Table 6.10   Selected financial indicators, 1999–2006 (kyat million)

	Demand deposits	Time, savings and foreign currency deposits	Money and quasi money (M2)
1999	72,707	216,549	562,224
2000	119,746	335,574	800,542
2001	206,349	450,560	1,151,713
2002	290,520	541,307	1,550,778
2003	82,948	386,298	1,572,402
2004	139,880	594,169	2,081,824
2005	209,324	697,736	2,651,111
2006*	233,765	699,953	2,772,768

* as at end of February
Sources: International Monetary Fund, 2006. *International Financial Statistics*, various issues, International Monetary Fund, Washington, DC. Myanmar Central Statistical Office (MCSO), 2006. *Selected Monthly Indicators*, Myanmar Central Statistical Office, Rangoon. Available from http://www.csostat.gov.mm

and inadequate liquidity support to the banks by the CBM was negated overwhelmingly by the imposition of 'withdrawal limits' on depositors that escalated into an outright denial of access for depositors to their money. Even worse, loans were 'recalled' with little consideration given to the capacity to repay. More serious breaches of trust in banking would be difficult to imagine. With a full-scale banking crisis in play, the usual symptoms of such events followed: bank closures and insolvencies, a flight to cash, the cessation of lending, the stopping of remittances and transfers, and other maladies destructive of monetary institutions. By mid 2003, the private banks had essentially ceased to function. In 2004, selected banks reopened, some of the largest closed completely (including the Asia Wealth Bank and the Myanmar Mayflower Bank, then the largest and third largest respectively of Myanmar's private banks) and an anaemic recovery began.

Demand as well as less liquid deposits have bounced back, though the former are still below the levels of late 2002 (Table 6.10). Taken together, in February 2006, total bank deposits of K933,718 million were a mere 33.7 per cent of the total money supply (M2)—indicating that the State remains by far the dominant actor in Myanmar's financial sector (see also Table 6.3).

Of course, the data in Table 6.10 can also be profitably employed once more to critique the SPDC's growth claims in recent years. For instance, the regime claimed that Myanmar's economy grew a vigorous 10.6 per cent in 2003, a year in which new lending to the private sector ceased, loans financing existing activities were recalled and all measures of private monetary assets declined dramatically. In short, if one was to believe Myanmar's official statistics, the country has been able to grow strongly not only without the increased use of energy and other 'real' factors of production, but seemingly without money.

Money laundering

The shadow of money laundering continues to linger over Myanmar's financial sector, even though the country has finally now been deemed a 'cooperative' jurisdiction with respect to money laundering by the

Financial Action Task Force (FATF). The FATF, an associate body of the Organisation for Economic Cooperation and Development (OECD), is the world's premier agency for dealing with money laundering globally. Myanmar had been named as a non-cooperating country in each of the FATF's annual reports since the organisation's inception in 1998, and was named again in 2006—the last country so designated.[13] In October, however, the FATF reported that, due to its 'good progress in implementing its anti-money laundering system', Myanmar would be removed from the 'non-cooperative' list.[14] At the same time, the FATF advised that Myanmar would continue to be monitored to ensure further progress, and that the country had been urged to 'enhance regulation of the financial sector...and to ensure that dealers in precious metals and precious stones follow anti-money laundering requirements'.[15]

Agriculture

Myanmar remains an overwhelmingly agricultural country. Agriculture accounts for about 57 per cent of Myanmar's GDP and engages more than 70 per cent of its labour force (FAO 2004:5). Nevertheless, for many years it has been a sector of profound neglect and routine exploitation by the government. Critical inputs such as fertiliser are unavailable to most farmers at prices they can afford, and more than 80 per cent of Myanmar's land under cultivation lacks irrigation of any form (Dapice 2003; EIU 2006:22). As noted earlier, credit from formal institutions is unavailable to most farmers in Myanmar, and at present less than 3 per cent of bank lending in Myanmar is extended to agriculture. Most of this is advanced by the state-owned Myanmar Agricultural Development Bank (MADB), which is inexorably 'decapitalising' in the face of continuing losses and Myanmar's chronically high inflation. Illogically, the private banks are forbidden to lend for farming (FAO 2004:13). Recent experiments in micro-finance under the auspices of the United Nations Development Programme (UNDP) show promise, but are at great risk from the lack of legal recognition accorded to them by Myanmar's government, as well as recent SPDC controls applying to the movements of UN agency and NGO staff (Turnell 2005).[16]

In 2003, Myanmar formally liberalised the trade in rice, internally and externally,[17] but in practice great interference by the State in the basic decisions taken by farmers—what, how and how much to produce—continued unabated. Of course, in many areas of Myanmar a final blow is the exaction of Myanmar's military forces, the *Tatmadaw*, forced by the country's strained finances to 'live off the land' (Vicary 2003, 2004). In recent years, the SPDC has adopted a number of programs designed to increase the amount of land under cultivation in Myanmar. Such efforts, which include the so-called 'summer paddy program', and various schemes designed to reclaim land in the Irrawaddy Delta, have invariably failed to achieve their desired outcomes because of the lack of the critical inputs noted above.[18] Farmers without sufficient fertiliser to prepare new fields, or without credit to allow the construction of dykes, fences and other land improvements, have been unable to make effective the exhortations for more extensive production (Okamoto et al. 2003; Thawnghmung 2004). In perhaps a sign that this problem has been recognised, while simultaneously pressing for greater scale in agricultural production, in December 2005, the SPDC announced it had signed a memorandum of understanding with Thailand that would allow Thai investors to cultivate some seven million hectares of vacant agricultural land in Myanmar. This author cannot but agree with the EIU on the venture, however, and its scepticism that 'it remains to be seen whether the junta will have any success in attracting significant Thai investment into the sector' (2006:22).

The end result of all of these supply-side problems (just some of which are noted above) is that Myanmar's agricultural sector, once the jewel of its economy (the famed rice bowl of Asia) is operating well below potential. According to the FAO (2004:28), 'the available data appears [sic] to indicate stagnant (agricultural) productivity growth and rising rural poverty since the mid 1990s.' The UNDP (2006:7) concurs with this bleak assessment, noting that the average household in Myanmar spends in excess of two-thirds of its income on food, a ratio that is high by international standards and the highest in the region. Meanwhile, a survey conducted by UNICEF in 2003 found that one-third of children in Myanmar suffered from malnutrition.[19]

Conclusion

In 2006, Myanmar's possession and exploitation of prized natural resources continued to flatter the appearance of the country's economic circumstances. Behind this façade, however, is a narrative of chronic failure that is the consequence of a political economy that is yet to create the institutions necessary for long-term economic development. Such institutions, which include effective property rights, freedom to contract and a modicum of macroeconomic stability, are created out of domestic constituencies possessing incentives to bring about change. The economic rents that are accruing from Myanmar's offshore energy deposits could further weaken these constituencies. Optimism with regard to Myanmar's economy accordingly must remain, for the moment, suspended.

Notes

1 Quoted in *The Myanmar Times*, 16(305), 20–26 February 2006.
2 As noted below, such revenue should further increase from June 2006 as a consequence once more of a revision in the 'dutiable' exchange rate applied to imports.
3 The author is grateful to Trevor Wilson for this point.
4 The minister's plea was reported in *The Myanmar Times* 2006.
5 Technically, the kyat is fixed to the IMF's 'Special Drawing Rights' at a rate of K1:SDR8.5085—which yields a more or less constant K6:US$1 (IMF 2006).
6 The author can confirm that the US dollar also remains the favoured medium through which larger Burmese businesses continue to conduct their activities.
7 The IMF's Article IV consultations, which take place (usually annually) with all of the IMF's member countries, are concerned with appraising members' economic, financial and exchange rate policies. The IMF's findings and recommendations are subsequently submitted to the governments concerned. The latest Article IV consultations with Myanmar took place in October 2006. A press release giving a brief outline of the talks is available at http://www. imf.org/external/np/sec/pr/2006/pr06216.htm
8 This figure, based on official Burmese data, is lower than that suggested by Thai import data. Accordingly, it probably understates Myanmar's gas exports in 2004.

9 This figure for agricultural investment, which is consistent with other sources, was rather surprisingly reported in the Yangon-based *Weekly Eleven News* in December 2005. The report was reproduced the same month in the online edition of *The Irrawaddy*. Available from http://www.irrawaddy.org

10 This figure is consistent with that estimated by British Petroleum (BP) in a review of global gas reserves: see *The Economic Times* 2006.

11 This saga was reported widely in the press at the time, representative of which are the accounts of *The Irrawaddy* magazine.

12 'Daewoo unlocks Burma's giant gas reserve', *The Irrawaddy*, online edition, January 2006.

13 The FATF's annual report for 2006 can be found at http://www.fatf-gafi. org/

14 The FATF press release announcing Myanmar's removal from the 'non-cooperative' list can be found at http://www.fatf-gafi.org/ dataoecd/45/25/37546739.doc

15 http://www.fatf-gafi.org/dataoecd/45/25/37546739.doc

16 See Trevor Wilson's chapter in this volume on Myanmar's foreign relations for an explanation of these restrictions.

17 See Ikuko Okamoto's chapter in this volume on the liberalisation of Myanmar's rice-trading system.

18 For more on the failure of these schemes, see the chapter by Mary Callahan in this volume on developments in the military, and Ikuko Okamato's chapter noted above.

19 UNICEF report cited in UNDP (2006:7). Other recent accounts attesting to the increasingly desperate circumstances in rural Myanmar include FAO 2004; Aung Din Taylor 2002; and Dapice 2003.

References

Asian Development Bank, 2004. *Asian Development Outlook 2004*, Asian Development Bank, Manila.

——, 2005. *Asian Development Outlook 2005*, Asian Development Bank, Manila.

Bradford, W., 2004. '*Fiant fruges?* Burma's *sui generis* growth experience', *Burma Economic Watch*, No.2:6–14.

Dapice, D., 2003. 'Current economic conditions in Myanmar and options for sustainable growth', Global Development and Environment Institute Working Paper No.03-04, Tufts University, Boston.

Economist Intelligence Unit, 2004. *Burma (Myanmar): country profile*, Economist Intelligence Unit, London.

——, 2005. *Burma (Myanmar): country profile*, Economist Intelligence Unit, London.

——, 2006. *Burma (Myanmar): country report*, May, Economist Intelligence Unit, London.

Financial Action Task Force, 2006. *Annual Report*, Financial Action Task Force. Available from http://fatf-gafi.org/

——, 2006. 'Chairman's summary', *Vancouver Plenary*, Financial Action Task Force. Available from http://www.fatf-gafi.org/dataoecd/45/25/37546739.doc

Food and Agriculture Organization, 2004. 'Myanmar: agricultural sector review and investment strategy', Working Paper 9: rural finance, Food and Agriculture Organization, Rome.

Fullbrook, D., 2006. 'Resource-hungry China to devour more of Burma's gas and oil industry', *The Irrawaddy*, 1 February 2006, online edition. Available from http://www.irrawadday.org (accessed 15 January 2007).

International Monetary Fund, 2006. *International Financial Statistics*, various issues, International Monetary Fund, Washington, DC.

Lwin, Y., 2006. 'Border trade in kyats helps tax earnings', *The Myanmar Times*, 17(321), 19–25 June 2006.

Miles, M.A., O'Grady, M.A. and Holmes, K.R., 2006. *2006 Index of Economic Freedom: the link between economic opportunity and prosperity*, Heritage Foundation, Washington, DC.

Myanmar Central Statistical Office, 2006. *Selected Monthly Indicators*, Myanmar Central Statistical Office, Rangoon. Available from http://www.csostat.gov.mm (accessed 15 January 2007).

Oil and Gas Journal Online, 2006a. 'Daewoo, partners make gas strike off Myanmar', *Oil and Gas Journal Online*, 23 June 2006. Available from http://www.ogj.com (accessed 15 January 2007).

——, 2006b. 'Myanmar steps up gas production from Yetagun field', *Oil and Gas Journal Online*, 14 August 2006. Available from http://www.ogj.com (accessed 15 January 2007).

Okamoto, I., Kurita, K., Kurosaki, T. and Fujita, K., 2003. Rich periphery, poor center: Myanmar's rural economy under partial transition to market economy, mimeo., October, Institute of Economic Research, Hitotsubashi University, Tokyo.

Taylor, D.A.D., 2002. 'Signs of distress: observations on agriculture, poverty, and the environment in Myanmar', paper delivered to the Conference on Burma: Reconciliation in Myanmar and the Crises of Change, 22 November 2002, School of Advanced International Affairs, Johns Hopkins University, Washington, DC.

Thawnghmung, A.M., 2004. *Behind the Teak Curtain: authoritarianism, agricultural policies and political legitimacy in rural Burma*, Kegan Paul, London.

The Economic Times (India), 2006. 'India, Myanmar sign gas supply deal', *The Economic Times* (India), 10 March 2006.

The Irrawaddy 2006. 'Daewoo unlocks Burma's giant gas reserve', *The Irrawaddy* online edition, January. Available from http://www.irrawaddy.org (accessed 15 January 2007).

The Myanmar Times, 2006. 'Govt aims for 5pc inflation', *The Myanmar Times*, 16(315), 8–14 May.

Turnell, S.R., 2003. 'Myanmar's banking crisis', *ASEAN Economic Bulletin*, 20(3):272–82.

——, 2004. 'Burma bank update', *Burma Economic Watch*, 1:19–27.

——, 2005. 'A survey of microfinance institutions in Burma', *Burma Economic Watch*, 1:29–35.

——, 2006. 'Burma's economy 2004: crisis masking stagnation', in T. Wilson (ed.), *Myanmar's Long Road to National Reconciliation*, Institute of Southeast Asian Studies, Singapore:77–97.

United Nations Development Programme, 2006. *Human Development Initiative, Myanmar: report of independent assessment mission, 4–29 July 2005*, United Nations Development Programme, New York.

Vicary, A.M., 2003. 'Economic non-viability, hunger and migration: the case of Mawchi Township', *Burma Economic Watch*, 1:1–18.

——, 2004. 'The state's incentive structure in Myanmar's sugar sector and inflated official data: a case study of the industry in Pegu Division', *Burma Economic Watch*, 2:15–28.

Xinhua News, 2006. 'More gas found in Myanmar offshore', *Xinhua News*, 6 March.

7 Transforming Myanmar's rice marketing

Ikuko Okamoto

Creating a rice[1] marketing system to serve the national interest has been one of the central policy issues for the Myanmar government since independence. It is no exaggeration to say that agricultural policy in Myanmar has been synonymous with rice policy.

Under the socialist government, a comprehensive system of controls over rice marketing was established for the first time, which introduced a rice rationing system for consumers along with a compulsory delivery system for procuring paddy directly from farmers to support the rationing system. At the same time, the exportation of rice became a state monopoly and served as the regime's main source of foreign exchange. These were the pillars of agricultural policies in the socialist period.[2]

The liberalisation of agricultural marketing in Myanmar began in the late 1980s, starting with the domestic agricultural market in 1987. This move signalled the start of Myanmar's transition to a market economy. In 1988, the ban on the private export of agricultural produce was lifted, and thereafter the marketing of some crops enjoyed full liberalisation. Rice marketing, however, which was originally the main target of agricultural reform, remained under state control. The rice rationing system was maintained for public servants, and the paddy procurement system, which had been terminated in 1987, was revived in 1989. Further, rice exporting continued to be a government monopoly. This sequence of reform can be called the first liberalisation.

In April 2003, 16 years after the first liberalisation, another liberalisation of rice marketing was suddenly announced. Under this second liberalisation, the rice rationing system for public servants and the paddy procurement system were abolished. Initially, the private export of rice was also incorporated into the reform plan; however, this part of the plan was not implemented when abolition of the rice rationing system was announced in January 2004. The aftermath of the second liberalisation also shows that the government is still not ready to undertake full-scale rice export deregulation.

The rationale of these two liberalisations was not found in improving efficiency of the rice marketing sector. Rather it was to keep the rice price at a low level, mainly for the sake of political stability. By examining the transformation of Myanmar's rice marketing closely, this paper attempts to show how this characteristic of policy and liberalisation influenced the development process of the rice marketing sector—state and private—in Myanmar, as well as the overall economy.

The state marketing system after the first liberalisation

The procurement of paddy

Under the first liberalisation, with the decrease in the volume of rationed rice compared with the socialist period, the paddy procurement system that supplied the rice was scaled back. Initially, it was planned to collect paddy for the rationed rice supplies in the form of land revenue from farmers and commercial taxes from traders. The new collection system, however, became caught up in the political upheaval of 1988 and did not function well, with the result that the amount collected fell far short of requirements. The next year, the government revived the paddy procurement system, which had a strong institutional base under the socialist government.

In a determined effort to achieve its procurement goals, the government sought to placate farmers by reducing the pressure on them. A procurement quota was set for paddy produced in the monsoon season (monsoon paddy), but it was decreased to 10–12 baskets[3] per acre from the 30–40 baskets per acre of the socialist period. This meant that the volume of rice procured by the government as a share of total rice production decreased by one-third after liberalisation (Table 7.1).

Despite the official assertion that the burden of the paddy procurement system on farmers was eased, there were various problems in the procurement process. First, the amount procured was fixed on a per acre basis; thus farmers with lower productivity or less marketable surplus were at a disadvantage. Unlike the system in the socialist period, which absorbed the farmers' entire marketable surplus, the new system had the merit of inducing farmers to increase production. The disadvantage, however, was that it did not reflect the disparity in the

Figure 7.1 Changes in procurement and farm-gate prices, 1999–2002

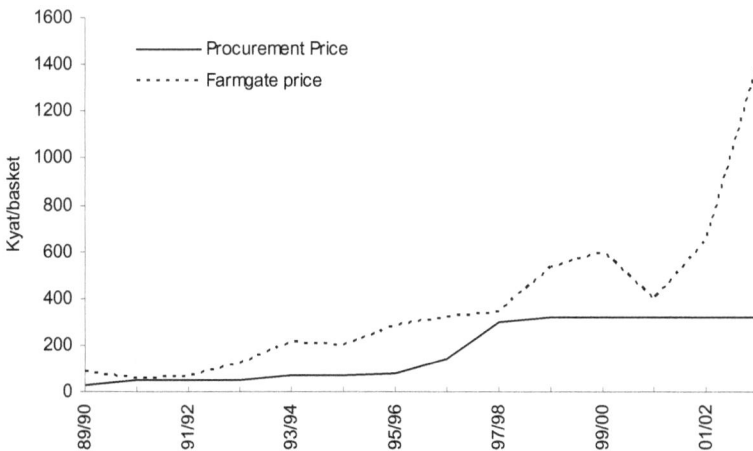

Sources: Author's survey. Myanmar Agricultural Produce Trading, 1991. *MAPT in Figures* (in Burmese). Takahashi, A., 2000. *Myanmar's Village Economy in Transition: changing rural life in a market-oriented economy*, University of Tokyo Press, Tokyo.

Table 7.1   Estimated volume of domestically marketed rice (in terms of paddy), 1970–2001

	A Production	Procurement b	Ratio (%) b/A	Deduction Seed	Waste	Home consumption	C-A-B Marketed volume	Ratio (%) C/A	Milled rice	D Export Converted to paddy d	Share in the procurement (%) d/b	Share in production (%) d/A
1971–72	8,189	2,245	27.4	514	514	7,528	(1,585)	(31.9)	831	1,240	55.3	15.1
1976–77	9,335	2,889	30.9	524	524	7,538	(2,140)	(22.9)	646	964	33.4	10.3
1980–81	13,340	4,259	31.9	530	530	7,384	637	4.8	703	1,049	24.6	7.9
1985–86	14,341	4,156	29.0	506	506	7,354	1,818	12.7	594	887	21.3	6.2
1986–87	14,150	4,263	30.1	500	500	7,363	1,523	10.8	604	901	21.1	6.4
1987–88	13,658	564	4.1	482	482	7,402	4,728	34.6	320	478	84.7	3.5
1988–89	13,186	1,672	12.7	494	494	7,447	3,080	23.4	48	72	4.3	0.5
1989–90	13,826	1,482	10.7	504	504	7,551	3,785	27.4	169	252	17.0	1.8
1990–91	13,748	1,851	13.5	511	511	7,579	3,296	24.0	134	200	10.8	1.5
1991–92	12,993	2,095	16.1	499	499	7,589	2,312	17.8	183	273	13.0	2.1
1992–93	14,603	2,222	15.2	530	530	7,648	3,672	25.1	199	297	13.4	2.0
1993–94	15,500	1,939	12.5	587	587	7,694	4,693	30.3	261	390	20.1	2.5
1994–95	17,908	2,034	11.4	613	613	7,737	6,911	38.6	1,041	1,554	76.4	8.7
1995–96	17,669	1,934	10.9	634	634	7,772	6,695	37.9	354	528	27.3	3.0

1996–97	17,397	1,522	8.7	607	607	7,810	6,852	39.4	93	139	9.1	0.8
1997–98	16,391	1,601	9.8	597	597	7,829	5,765	35.2	28	42	2.6	0.3
1998–99	16,808	2,200	13.1	607	607	7,869	5,524	32.9	120	179	8.1	1.1
1999–2000	20,159	2,212	11.0	649	649	7,908	8,741	43.4	69	103	4.7	0.5
2000–01	21,359	2,126	10.0	657	657	7,948	9,972	46.7	257	384	18.0	1.8

Notes: Seed and waste are assumed to be two baskets per acre. Home consumption is calculated as the number of households multiplied by 5.5 (the average number of people per household in 1999) by 15 baskets. For 1998–99 to 1999–2000, since the data for farm households were not available, it was estimated using the average increase rate of households. Export includes white and broken rice. The paddy conversion rate for export is assumed to be 67 per cent of households.

Sources: Government of Union of Myanmar. *Review of the Financial, Economic and Social Conditions (REFS)*, various issues, Yangon. Government of Union of Myanmar, 2001. *Myanmar Agricultural Statistics (1989–90 to 1990–2000)*. Central Statistical Office, 1991. *Myanmar Statistical Yearbook*, Central Statistical Office, Yangon. Central Statistical Office, 1998. *Myanmar Statistical Yearbook*, Central Statistical Office, Yangon. Central Statistical Office, 2001. *Myanmar Statistical Yearbook*, Central Statistical Office, Yangon. Tin Htut Oo and Kudo, T.T., 2003. *Agro-Based Industry in Myanmar: prospects and challenges*, Institute of Developing Economies, Japan External Trade Organization, Chiba:Table 15.

productivity of individual farmers or take into consideration reasons for fluctuations in yield, such as weather.

Second, (Figure 7.1) there was a persistent disparity between the procurement and market prices. Even though it was paid in advance to meet some of the cultivation costs, the procurement price paid under this system was kept at 40–60 per cent of the prevailing free-market price. This suggests that the real burden on farmers was not lessened to the extent that the government asserted.

Finally, there was the problem of the quality of procured paddy. In response to the government's low procurement price, farmers tended to deliver to the depots their lower-quality paddy (such as that which was not fully dried or had been intentionally mixed with foreign matter) and sold their better paddy on the free market. Another factor affecting quality was that paddy delivered to the depot was supposed to be separated into varieties, but in practice this separation was loosely controlled and different varieties became intermixed. Thus good-quality paddy could become mixed with poor-quality paddy, leading to a lower grade of milled rice. Subsequently, the quality of procured paddy became a big problem, much as it had been in the socialist period (Takahashi 1992).

The milling of officially procured paddy

The paddy collected from farmers was milled either at rice mills owned by the government's Myanmar Agricultural Produce Trading (MAPT) enterprise or private mills contracted to MAPT. As of 2000–01, MAPT owned 68 mills, mainly in the major rice-producing areas. Most of these were constructed in the 1980s with official development assistance from Japan or international organisations. Many of MAPT's mills were large scale with a capacity of 100 tonnes of milled rice a day, while most private mills had a capacity of less than 50 tonnes a day. There was far more paddy procured than MAPT could handle at its own mills, so it contracted with private mills. After liberalisation began in 1987, the share for MAPT mills was only 32 per cent, on average, indicating the government's great dependency on private mills (Table 7.2).

Table 7.2   Changes in milled rice by MAPT-owned and MAPT-
             contracted mills , 1988–2001

Fiscal year	Procurement (basket)	MAPT mills Paddy (basket)	MAPT mills Milled rice (tonne)	MAPT contracted mills Paddy (basket)	MAPT contracted mills Milled rice (tonne)	Share of MAPT mills (%)	Share of milled rice in the total procured amount (%)
1988–89	85.10	14.70	0.18	46.10	0.58	24.18	71.4
1989–90	63.00	19.10	0.24	59.20	0.78	24.39	124.3
1990–91	72.10	19.60	0.24	40.40	0.53	32.67	83.2
1991–92	74.70	20.30	0.25	45.90	0.59	30.66	88.6
1992–93	76.50	25.00	0.31	57.70	0.75	30.23	108.1
1993–94	92.30	27.00	0.34	50.90	0.67	34.66	84.4
1994–95	97.30	32.10	0.40	76.50	0.97	29.56	111.6
1995–96	92.90	27.40	0.35	67.10	0.85	28.99	101.7
1996–97	73.00	22.60	0.28	49.90	0.65	31.17	99.3
1997–98	44.70	21.70	0.27	37.20	0.48	36.84	131.8
1998–99	105.30	26.20	0.33	46.00	0.61	36.29	68.6
1999–2000	105.83	30.90	0.38	53.30	0.69	36.70	79.6
2000–01	101.74	28.10	0.35	51.80	0.67	35.17	78.5

Source: Tin Htut Oo and Kudo, T.T., 2003. *Agro-Based Industry in Myanmar: prospects and challenges*, Institute of Developing Economies, Japan External Trade Organization, Chiba:114. MAPT documents

One reason for the high dependency on private rice mills after liberalisation, even with the decrease in the volume of procured rice, was the run-down condition of MAPT's mills. These facilities could not be maintained or repaired after the halt of overseas development assistance after 1988. Also, the chronic shortage of electricity greatly lowered their rate of operation, as most of MAPT's mills were powered by electricity. Some mills operated only six to 10 hours a day because of blackouts, although they had 24-hour operating capacity.

Rice rationing for the budget group

The rice rationing system targeting general consumers was abolished with the first liberalisation, and the system was limited to targeting the so-called Budget Group, which consisted of public servants and military personnel. The government could not abandon the whole rationing system in the midst of the tense political situation in 1988; it had to be maintained at least for the public servants to secure the political base of the regime. The number of people targeted for rice rations reduced the volume of rationed rice to 6–800,000 tonnes in normal years. This was a decrease of one million tonnes compared with the volume rationed in the socialist period (Table 7.3).

Table 7.3   Changes in volume of rationed rice, 1980–2002 ('000 tonnes)

	Procurement volume of paddy	Rationed rice Rice	Converted in paddy	Share of rationed volume in the procurement (%)
1980–81	4,259	1,618	3,236	76.0
1983–84	4,145	1,709	3,418	82.5
1987–88	564	574	1,148	203.5
1988–89	1,672	556	1,112	66.5
1989–90	1,482	869	1,738	117.2
1990–91	1,851	751	1,502	81.1
1991–92	2,095	616	1,232	58.8
1992–93	2,222	770	1,540	69.3
1993–94	1,939	711	1,421	73.3
1994–95	2,034	744	1,487	73.1
1995–96	1,934	769	1,539	79.5
1996–97	1,522	822	1,643	108.0
1997–98	1,601	773	1,546	96.6
1998–99	2,200	668	1,336	60.7
1999–2000	2,212	616	1,232	55.7
2000–01	2,126	585	1,169	55.0
2001–02	2,119	569	1,137	53.7

Note: The conversion rate from rice to paddy is 50 per cent.
Sources: Government of Union of Myanmar. *Review of the Financial, Economic and Social Conditions (REFS)*, various issues, Yangon. Ko Ko Gyi, 1994. Public and private marketing channels for food grains situation and improvements needed, Paper presented at the FAO/AFMA/Myanmar Training Workshop, 21–25 November:Table 5. MAPT 2003:222–3. MAPT documents

The quality problem of procured paddy, pointed out earlier, also affected the rice rationing system. Although this system was beneficial for recipients in terms of volume and price,[4] it was not sufficient to overcome the inferior quality of the rice, which led recipients to sell it to traders as feed for livestock rather than consume it at home. Consequently, the rice rationing system no longer worked as a benefit for its recipients, as the government originally intended.

Rice exports

Rice exporting remained the monopoly of the government, and the main agency for this was MAPT, even after 1988. The government monopoly on rice exporting was utilised as a measure to control the price of rice for the general consumer, who was excluded from the rice ration system after the first liberalisation. A general deregulation of private exporting was announced only two months after the peak of the democracy movement. In other words, the government wanted to maintain a stable rice price for general consumers for fear of further instability, and it regarded the preservation of its monopoly on rice exporting as one means to this end. Consequently, private rice exporting was not allowed.

The government's priority was on securing rice for rationing, and only the rice remaining in government hands after rationing was released for export. Consequently, only an extremely small amount of rice was exported compared with during the socialist period (see Table 7.1).

Due to the inferior quality of procured paddy, the destinations for exported Burmese rice were limited. A breakdown of Myanmar's rice exports (Table 7.4) shows that most went to South Asia, Africa and Southeast Asia, representing a large proportion of the world's low-income countries where demand for low-quality rice was high. But Burmese rice has failed to generate stable export demand because of its export regime, which depends greatly on the state marketing sector.

Through its monopoly over rice exports, however, the government was successful in separating the domestic and international markets, which led to a huge disparity between the domestic and international

Table 7.4 Direction of Myanmar's rice exports, 1990–2001 (percentage by volume)

	1990–91	1991–92	1992–93	1993–94	1994–95	1995–96	1996–97	1997–98	1998–99	1999–2000	2000–01
Southeast Asia	11.2	25.7	2.0	6.1	61.0	73.7	50.5	3.6	55.0	36.4	18.3
South Asia	49.3	26.2	37.7	18.8	9.5	7.3	21.5	96.4	15.8	41.8	69.3
Rest of Asia	-	4.9	-	3.1	-	5.1	-	-	0.8	-	-
Africa	29.9	43.2	57.3	66.7	26.5	6.5	26.9	-	25.8	-	10.0
Middle East	2.2	-	3.0	1.1	-	-	-	-	-	-	-
North and South America	7.5	-	-	-	1.4	7.3	-	-	-	-	-
Europe	-	-	-	4.2	1.5	-	1.1	-	2.5	21.8	2.4
Total	100.0	100.0	100.0	100.0	100.0	100.0	100.0	100.0	100.0	100.0	100.0

Sources: Central Statistical Office, 1997. *Myanmar Statistical Yearbook*, Central Statistical Office, Yangon.;Central Statistical Office, 2001. *Myanmar Statistical Yearbook*, Central Statistical Office, Yangon.

prices for rice. The domestic rice price at the free-market, foreign-exchange rate was 60 per cent of the international price, on average, after the first liberalisation. It even fell to 40 per cent of the international price when the domestic price collapsed in 2000–01. The international price for rice has been trending downwards in the past two decades, but the Myanmar government has kept the price of domestic rice well below even the declining international level.

The private rice marketing sector after the first liberalisation

Development of the private marketing sector

The rice ration system and its supporting procurement system were scaled back after the first liberalisation and the private sector came to play a larger role in supplying rice to the general consumer. The first liberalisation abolished the restrictions on private millers and traders as well as the geographical restrictions on rice trading that existed in the socialist period.

The shrinking of the state marketing sector along with the government's policy in the 1990s of raising rice production brought a steady increase in the volume of rice on the free market. The volume reached 30–40 per cent of total production by the end of the 1990s (see Table 7.1)—and rice came to be marketed widely in the country, supported partly by the development of transport infrastructure.

Responding to the increase of the marketable surplus of rice, private rice millers and traders actively entered the market. The number of private rice mills increased throughout the 1990s; there was a particularly sharp rise in the number of small mills in the villages (often called huller mills, the capacity of which was below 15 tonnes a day). The exact number of these small rice mills is not available, but there are normally one to five of them in each village tract. Assuming that there are two rice mills in a village tract in the major rice-producing areas (for example, Ayeyarwaddy, Bago, Yangon and Mandalay Divisions and Mon State), the total number of these small mills could be as high as

14,240. Needless to say, this is a rather conservative estimate. Most of these mills handle paddy for home consumption in the villages, while some engage in milling for sale on the free market.

As well as the rice millers, a large number of traders entered the rice market. According to the author's survey of 47 wholesalers in eight major rice markets in 2002, 39 wholesalers (84.8 per cent) began rice trading after liberalisation in 1987; only five (10.9 per cent) were doing so before then. By far the greater share of rice traders entered the market after liberalisation. The formation of marketing networks over wide areas of the country as well as the increase in the volume of marketed rice produced by farmers encouraged the entry of traders, especially in the late 1990s.

Problems the private rice marketing sector has faced

As pointed out earlier, there was a remarkable increase in the number of small rice mills in rural areas in Myanmar. In contrast, however, medium and large-scale rice mills (mills with a milling capacity of more than 16 tonnes of rice a day) decreased in number. Changes in the number of MAPT-registered mid-size and large rice mills shows a sharp decrease in the number of these rice mills in only two years (Table 7.5).[5]

The majority of these big rice mills were established during the British colonial period or the socialist period. Those opened during the colonial period played a primary role in making Myanmar one of the giant rice exporters of the world. When rice exporting became a government monopoly in the early 1960s, however, these mills were required to mill the government-procured paddy at the official fixed rate—although they were not nationalised in the strict sense. After the first liberalisation in 1987, these medium and large-scale mills were also allowed to operate in the private rice market, but business was not easy.

One reason for this is that the mills failed to utilise their capacity fully because of the scale of their facilities. Behind this lies the decreasing demand for milling at medium and large-scale rice mills. The rapid increase in the number of small mills in the villages after the first liberalisation reduced the need to transport paddy to the distant big mills and their rate of operation declined. Before liberalisation, the rice for rural household consumption was milled primarily at big mills located in towns. During the 1990s, however, this rice came to be processed mostly at the newly established village mills, and the big town mills lost business. Given the downward trend in the demand for milling at medium and large-scale rice mills, and in an effort to raise their rate of operation, some of these big mills shifted from specialising in custom milling and started normal milling, whereby the mills bought and milled paddy at their own expense and then sold the rice themselves. This was another indication of the unfavourable business conditions facing the big rice mills.

A second problem for medium and large-scale rice mills was that the milling of MAPT paddy often became a burden—financially and physically. Even though MAPT bore the cost of labour for the milling of its paddy, big mills contracted by MAPT still often found that milling for the organisation did not pay: the milling fee paid to contracted

Table 7.5   Number of private mills registered with MAPT, 1998–2001

State/Division	1998–99	2000–01	1998–99 (%)	2000–01 (%)
Ayeyarwaddy	489	369	47.2	53.7
Bago	208	133	20.1	19.4
Yangon	123	69	11.9	10.0
Mon	66	32	6.4	4.7
Rakhine	5	4	0.5	0.6
Sagaing	49	41	4.7	6.0
Mandalay	68	11	6.6	1.6
Magwe	6	0	0.6	0.0
Kachin	8	10	0.8	1.5
Tanintaryi	2	2	0.2	0.3
Kaya	11	16	1.1	2.3
Total	1,035	687	100.0	100.0

Sources: MAPT documents, Tin Htut Oo and Kudo, T.T., 2003. *Agro-Based Industry in Myanmar: prospects and challenges*, Institute of Developing Economies, Japan External Trade Organization, Chiba:Annex 7.

private mills was one-half to one-third of the prevailing free-market rate. For example, in 1998–99, the market milling fee was 20–30 kyats a basket, while MAPT paid only 10 kyats a basket. This meant that the farmers and the private millers were burdened by the rice rationing system. Further, the mills also needed to handle all the cumbersome procedures to abide by the requirements that MAPT prescribed. There were also cases where MAPT required these mills to store paddy or milled rice without payment. All these difficulties made the big rice mills reluctant to contract with MAPT. Table 7.6 shows the change in the number of mills contracted by MAPT to mill government-procured paddy. This number has been declining in the past decade, which can be interpreted as reflecting the general reluctance of private rice mills to contract with MAPT.

Finally, the biggest problem facing the medium and large-scale mills was the dilapidated condition of their facilities and equipment. Important parts of these mills, such as engines, have been in use since

Table 7.6   Number of private mills contracted for procurement of paddy, 1999–2001

Division/state	1991–92	1995–96	1998–99	2000–01
Ayeyarwaddy	220	208	144	138
Bago	173	136	100	51
Yangon	75	61	49	38
Mon	40	37	30	43
Rakhine	19	15	12	0
Sagaing	101	81	78	68
Mandalay	56	66	38	39
Magwe	32	19	20	15
Kachin	20	25	14	14
Tanintharyi	19	14	12	17
Kayin	9	1	1	0
Kaya	1	1	1	0
Total	765	664	499	423

Source: MAPT documents, Tin Htut Oo and Kudo, T.T., 2003. *Agro-Based Industry in Myanmar: prospects and challenges*, Institute of Developing Economies, Japan External Trade Organization, Chiba:Annex 5.

the 1930s; the most recent are from the 1960s. The cost of running and maintaining these old, second-hand mills with their worn-out equipment can be very high, but no support for maintenance or efficiency improvements has been forthcoming from the government, despite its dependence on the big mills for milling state-procured paddy.

In the view of most of the owners of the big rice mills, any substantial investment to upgrade facilities and improve quality will not pay, given that the market is still dominated by trading in medium and low-quality rice. Replacing their steam engines with electric motors would in all likelihood lower their rate of operation because of the chronic shortage of electricity. The limited supply of spare parts for reasonable prices and of sufficient quality has also detracted from the willingness of millers to undertake new investment. The majority of medium and large-scale millers say that they are ready to undertake new investment once private rice exporting is allowed and the market for high-quality rice expands. This clearly indicates that the present condition of Myanmar's rice market, characterised by government restrictions on exporting and the dominance of low and medium-quality rice, has narrowed the business opportunities for big rice millers, and this in turn has narrowed their business perspective.

The first liberalisation gave rice traders the freedom to deal in the domestic rice market, and this new market environment encouraged the entry of new rice traders. This freedom was, however, granted only on the condition that their dealings did not jeopardise the government's rice policy. Herein lay the main characteristic of the first liberalisation: rice traders were not entirely free from government intervention, which introduced an element of constant unpredictability into the sector.

The Myanmar government tended to intervene in the domestic rice market in three situations. One was when rice transactions were made with remote regions. In general, after the first liberalisation, there were no longer any restrictions on the marketing of rice over a wide area of the country; however, transactions with some remote regions bordering neighbouring countries were an exception. These regions were Shan, Chin and Rakhine States and Tanintharyi (Tennasserim) Division. For any rice transactions with these regions, permission from the local authorities was necessary. In some cases, the monthly quota for the volume of rice to be transacted was prescribed by the authorities. The ostensible rationale for this regulation was, of course, to keep the domestic rice price stable. With Myanmar's domestic rice price kept far below the international price, if sizeable amounts of rice were exported (even informally) to neighbouring countries, upward pressure on the domestic rice price would inevitably follow. To prevent this, every effort was made to regulate strictly the volume of rice transacted with these remote regions. This regulation, however, made the people in these regions, which are rice-deficit areas, pay a high price in relative terms for the rice they consumed.[6]

The second situation was when the volume of procured rice fell below the government's target. There was an unwritten rule, even when the harvest was normal, that traders could not buy paddy or rice from farmers who had not met their procurement quotas for that year. When procurement was not progressing well in an area, however, the government often prohibited all private sales of paddy or rice in that area. In the rice-deficit remote regions discussed above, the government generally did not permit such sales during the procurement season.

The third situation was when there was an abrupt rise in the rice price. The government was noticeably wary about depending on the private sector for the marketing of rice. Whenever the authorities judged that the rice price had gone above the level they could tolerate, orders were issued to start inspecting rice traders in various parts of the country, in rural and urban areas. As a result, compared with all other commodities, the rice market in Myanmar faces a much higher risk of sudden, unexpected intervention by the government. One rice trader commented, 'If you want to make a profit, don't go into rice trading; choose some other business.' Rice traders have to accept such interventions because the government maintains absolute vigilance against an unstable rice price.

The second liberalisation and its consequences

In the second liberalisation in April 2003, the government was pursuing three distinct policy agendas: one was to open up rice exports to the private sector; the second was to abolish the paddy procurement system; and the third was to retain the rice rationing system for the Budget Group by procuring rice from traders, not from farmers. In January 2004, however, private rice exporting was suddenly halted at the same time as the announcement of the abolition of the rice rationing system. Eventually, therefore, the second liberalisation encompassed only the liberalisation of the domestic rice-trading market.

What, then, was the background to and objective of the second liberalisation? First of all, the government's original objective was probably to earn a larger amount of foreign exchange through rice exporting. Evidence suggests that from the late 1990s, the government sought to export larger volumes of rice. Corroborating this, official data show that while the volume of ration rice was rather constant, the paddy volume procured from farmers was increasing (see Tables 7.1 and 7.3). This effort did not work as planned, and apparently the government decided to try another way, which was to earn more foreign exchange by increasing rice exports via the private sector. In order to give effect to this new approach, private rice millers and wholesalers were also allowed to become members of the Rice Trading Leading Committee, which the government had placed in charge of implementing the reform.

The original reform plan for exports was as follows: the government would open up rice exporting to private traders, by issuing export licences; the licences would enable the export of rice within a quota set annually by the government, with the government taking half of the foreign exchange earnings (it was equal to 45 per cent of the total earnings after the deduction of the 10 per cent export tax). In turn, the government would pay the marketing cost—equivalent to 45 per cent of exported rice—in local currency. After the second liberalisation, export licences were issued for 500,000 tonnes of rice, of which 270,000 tonnes were exported.

Secondly, the paddy procurement system was abolished as it no longer yielded the benefits to match the cost of retaining it. This was an indirect effect of the low rice price as a result of increased rice production. Because of the depressed domestic market price, rice production in general deteriorated in profitability. It became increasingly difficult to maintain the procurement system because the government had to procure paddy at a price even lower than the depressed market price. While it was true that after the first liberalisation the government had been able to preserve the procurement system by reducing the burden for farmers, the situation had reached a deadlock, though not in the form of the sort of farmer discontent observed in the mid 1980s. Worse still, even with greater government effort to procure rationed rice, recipients were finding little merit in it because of the generally low quality of this rice—a problem hampering the expansion of exports as well.

Added to this was MAPT's operating deficit, which had begun to widen again from the late 1990s. Soon after the first liberalisation, the procurement system deficit shrank remarkably when compared with that in the socialist period. According to MAPT, the deficit was 350 million kyats in 1986–87, which turned into a surplus of 310 million kyats by 1989–90. According to later MAPT documents, however, it appears that the deficit increased again from the mid 1990s, and especially at the end of the 1990s. It is possible that this increase was because of a rise in the procurement volume. The situation was beginning to resemble the adverse conditions for the rice sector at the end of the socialist period in the late 1980s.

As opposed to the liberalisation of rice exports and procurement of paddy/rice, the rice rationing system for the Budget Group of recipients was retained. This reflected the government's commitment to underpinning its political base. As the procurement system was to be abolished, procurement of the required amount of rice was arranged to take place through rice traders who were to be paid at the market price. Just before procurements were to start, however, the government realised that it would be difficult to cover the whole cost of rice procured at the market price and, in early 2004, it suddenly announced that the rice rationing system would also be abolished. To compensate government personnel for the loss of rationed rice, each person would receive a payment of 5,000 kyats a month.

This decision to abolish rice rationing and replace it with fixed cash payments had ramifications. If these payments were the only compensation for the cost of rice, it was likely that discontent would break out among public servants if the price of rice went up even by a small amount. This was a real concern for the authorities because there were signs that price increases would accompany export liberalisation. This possibility unnerved the government and it decided to freeze private rice exporting.[7] The reform plan was thus modified without discussion with the private sector. In the end, the stable supply of rice at a low price had top priority. The fundamental rationale of government rice policy prevailed over earning a larger amount of foreign exchange.

The significance of the second liberalisation in deregulating domestic rice marketing cannot be over-emphasised, as the domestic rice market was finally liberalised completely 42 years after the establishment of Myanmar's socialist government. MAPT, long the main organisation responsible for the rice procurement and rationing systems, lost its purpose for existing. Sizeable reduction of MAPT personnel began, and its rice mills were put up for sale.

The second liberalisation is expected to have three effects. First, the profitability of rice production and thus farmers' incomes are expected to improve. The sale of rice on the market is expected to increase by 10–20 per cent and rice production will become more market oriented. This will make farmers more concerned about the quality of rice they produce. In marginal rice-producing areas, where rice is grown mainly for home consumption, it is expected to lead to the reduction of rice purchased on the market. The second effect of this liberalisation will be a reduction in the number of situations in which the government can abruptly intervene in the market. This will reduce transaction costs for private rice traders.

The failure to open rice exporting to private traders after the second liberalisation was, however, a big set-back for the rice marketing sector in Myanmar from a mid to long-term perspective. Rice traders had been anticipating export deregulation and were greatly disappointed when it failed to take place. More than 20 export companies were set up in preparation for liberalisation, but these efforts were for nothing. The government's fickleness on the export issue has intensified rice traders' lack of confidence in the government, and traders are increasingly taking a risk-averse attitude towards new investment in facilities and the expansion of business. Without doubt, this is dampening the future outlook for the rice marketing sector in Myanmar.

Conclusion

The stable supply of rice at a low price continued to be the principal rationale of the rice marketing system in Myanmar even after the two liberalisations. The transition from comprehensive state control over rice marketing that began with the first liberalisation and continued with the second can be seen as an *ad hoc* transformation of the marketing system in response to the changing economic and political situation. It eventually took the form of gradual rice price deregulation. After the two liberalisations, Myanmar's rice-marketing system shifted from being one supported by the rice procurement and ration systems and export controls to one solely dependent on rice export controls to achieve the low rice price policy.

This policy orientation determined the development of the private rice marketing sector. The whole sector was allowed to develop only in the remaining sphere of the rice marketing sector and on condition that it did not jeopardise the stable supply of rice at a low price. This was the inevitable consequence of Myanmar's rice marketing policy. In the liberalisation process, however, the private rice marketing sector was able to achieve self-sustaining development. The government's policy to promote rice production and cut-backs in the volume of rice procurement increased the amount of rice sold in the market, which induced more traders to enter the rice-marketing business. This was a clear manifestation of the latent willingness of Myanmar's traders to grasp whatever small opportunities arose to increase profits, opportunities that had been closed for more than one-quarter of a century during the socialist period. The rice traders who expanded business while avoiding conflicts with the government rice policy were the ones who were able to survive during the 1990s.

By the end of the 1990s, however, the private rice marketing sector had reached a crossroads as the domestic rice market approached total saturation. This problem was most evident in the tough business conditions facing medium and large-scale rice millers. The worn-out state of their mills grew apace, but they could not risk venturing into new investments under the existing market structure where low and medium-quality rice was in greatest demand. Even in the milling of lower-quality rice, the big mills were losing out to the growing number of small-scale rice mills in the villages. Thus, by the time of the second liberalisation, medium and large-scale rice mills were facing a crisis in their operations.

What are the implications of this transformation of the rice sector in accordance with the low rice price policy to the development of Myanmar's national economy? The first implication is the poor prospects for the development of the rice industry. It cannot be denied that the commercial and processing industries of Myanmar's rice marketing sector continue to be the base of the rural economy. In neighbouring Thailand, rice millers turned to exporting and, with the accumulated capital, expanded their businesses to other industries with great success. In Myanmar, one would hope that the same scenario could play out for private rice traders and millers. In reality, however, there is little prospect that private rice exporting will be allowed in the near future. The present government is unlikely to change its rice policy, which prioritises a low price for the sake of political stability. Since export controls become the sole direct policy tool that the government has for keeping the price of rice low, it will remain reluctant to undertake any rapid deregulation of rice exports. This means that the private rice marketing sector will have to survive within the confines of the present domestic market, which limits demand largely to low and medium-quality rice. Thus the government's rice policy has again thwarted the development of Myanmar's rice industry and denied it the potential to stimulate growth in the economy as a whole.

The second implication, which could be more serious than the first, is the absence of a clear scenario to utilise the low rice price for development led by industrialisation (Fujita and Okamoto 2006). Generally speaking, the low rice price policy itself is not unique to Myanmar, and has been adopted in various developing countries, especially in the early stages of economic development. The purpose is to promote industrialisation using cheap labour, backed by the low price of rice. Any clear vision for this type of industrialisation has, however, been barely observed for Myanmar in the past 19 years. The low rice price policy has not gone beyond the purpose of maintaining the regime and it is likely to continue that way for some time.

Notes

1   In this chapter, rice means paddy and milled rice. When a distinction is necessary, the terms paddy or milled rice are used.
2   See Saito 1981, Takahashi 1992 and Tin Soe and Fisher 1990 for analyses of the procurement system in the socialist period.
3   One basket of paddy equals 20.9 kilograms.
4   Rice provisions were 25 kilograms a month for an unmarried public servant and 28kg for a married public servant. The price was kept at 21 per cent of the market price, on average, from 1988 to 2001.
5   During the author's survey in 1999 in a township in Yangon Division, there were 13 mid and large-scale rice mills, but only seven were operating. The other six had closed down.
6   According to the author's survey in 2001, the retail rice price in these remote regions was higher by 10–20 per cent compared with the average rice-deficit area in Upper Myanmar.
7   Along with rice, exports of chillies, onions, maize and sesame were also banned. This also reflected the high priority that the government put on self-sufficiency in important crops.

References

Central Statistical Office. *Statistical Yearbook*, various issues, Central Statistical Office, Yangon.

——, *Monthly Economic Indicators*, various issues, Central Statistical Office, Yangon.

Fujita, K., 2003. '90 Nendai Myanmar no Ine-Nikisakuka to Nogyo-Seisaku Noson-Kinyu: Irawaji-Kanku Ichi-Noson-Chosa-Jirei wo Chusin ni' [Policy-initiated expansion of summer rice and the constraints of rural credit in Myanmar in the 1990s: perspectives from a village study in Ayeyarwaddy Division], *Keizai Kenkyu*, 54(2):300–14.

Fujita, K. and Okamoto, I., 2006. *Agricultural policies and development of Myanmar's agricultural sector: an overview*, Discussion Paper Series No.63, Institute of Developing Economies, Chiba.

Government of Union of Myanmar, 2001. *Myanmar Agricultural Statistics (1989–90 to 1990–2000)*, Ministry of Agriculture and Irrigation, Yangon.

——, *Review of the Financial, Economic and Social Conditions (REFS)*, various issues, Ministry of Planning and Finance, Yangon.

Ko Ko Gyi, 1994. Public and private marketing channels for food grains situation and improvements needed, Paper presented at the FAO/AFMA/Myanmar Training Workshop, 21–25 November.

Kurosaki, T., Okamoto, I., Kurita, K. and Fujita, K., 2004. *Rich periphery, poor center: Myanmar's rural economy under partial transition to market economy*, COE Discussion Paper No.23, Institute of Economic Research, Hitotsubashi University.

Ministry of Agriculture and Irrigation, *Marketing Information Systems Price Bulletin*, various issues, Ministry of Agriculture and Irrigation.

——, 2000. *Agricultural Marketing in Myanmar*, Market Information Service Project TCP/MYA/8821, Ministry of Agriculture and Irrigation.

Myanmar Agricultural Produce Trading, 1991. *MAPT in Figures* (in Burmese), Myanmar Agricultural Produce Trading.

Saito, T., 1981. 'Farm household economy under paddy delivery system in contemporary Burma', *Developing Economies*, 19(4):367–97.

Takahashi, A., 1992. *Biruma Deruta no Beisaku-son: Shakaisugi Taiseika no Noson-Keizai* [A Rice Village in the Burma Delta: rural economy under the socialist regime], Institute of Developing Economies, Tokyo.

——, 2000. *Myanmar's Village Economy in Transition: changing rural life in a market-oriented economy*, University of Tokyo Press, Tokyo.

Tin Soe and Fisher, B.S., 1990. 'An economic analysis of Burmese rice policies', in M. Than and J. L. H. Tan (eds), *Myanmar Dilemmas and Options*, Institute of Southeast Asian Studies, Singapore:117–66.

Tin Htut Oo and Kudo, T.T., 2003. *Agro-Based Industry in Myanmar: prospects and challenges*, Institute of Developing Economies, Japan External Trade Organization, Chiba.

Acknowledgment

An earlier version of this chapter was published by the Institute of Developing Economies in Japan. See Okamoto, I., 2005. 'Transformation of the rice marketing system and Myanmar's transition to a market economy', discussion paper No. 43 (December), Institute of Developing Economies, Tokyo. Available online at http://www.ide. go.jp/English/Publish/Dp/pdf/043_okamoto.pdf.

8 Industrial zones in Burma and Burmese labour in Thailand

Guy Lubeigt

The military government's concerns with the industrialisation of Burma can be observed through the example of the development of satellite towns around Rangoon before the events of 1988 (Lubeigt 1989) and after them (Lubeigt 1993, 1994, 1995). The population surplus of downtown Rangoon and the squatters living around the pagodas and monastery compounds, who provided scores of demonstrators during the anti-socialist revolt, were expelled and forcibly resettled into the new townships created *ex nihilo* in far away paddy fields.[1] Potentially explosive crowds of Central Rangoon were dispersed to South and North Dagon, Shwepyitha and Hlaingthaya by a junta keen to get rid of these trouble-makers. Small private industries causing a nuisance in residential quarters subsequently were also resettled in special areas, which became *ipso facto* 'industrial zones'. Meanwhile, bigger enterprises, mostly textile joint ventures established with foreign capital under the 'market-oriented economy'—successor of the failed socialist economy—were set up in Mingaladon Township on the eastern side of the main Rangoon–Pegu (Bago) road.

Map 8.1   Industrial zones in Burma

The location of these factories was not chosen at random, as Mingaladon is the main cantonment of the capital. Military families could provide an excellent and obedient workforce for these enterprises. Meanwhile, the construction of factories since the beginning of the 1990s had been quite limited and insufficient to provide many job opportunities for the civilian population. Therefore, with a growing population in search of a living, the gap between unemployment and job opportunities increased dramatically. The newly designed industrial zones were intended to bridge this gap.

In 1995, the military government set up the Myanmar Industrial Development Committee to encourage the development of the industrial sector. Thus the creation of industrial zones on the territory of the union could also be presented as a government goodwill gesture to provide job opportunities to its unemployed citizens. Since 2003, the Burmese authorities claimed to have organised nearly 43,000 private industries scattered throughout the whole country (Ministry of Information 2006). All private factories (93 per cent of the industrial sector in 2005), however, are not set up within the industrial zones delimitated by the authorities. Of the 82,000 industries officially

Table 8.1   Development of the private sector in 2006

Subject	1988	2005	Progress
Private industries	26,872	42,707	15,835
Private industrialists	31,200	40,000	8,800
Business in industrial zones	-	8,463	8,463
Cottage industries	-	8,500	8,500
Number of industrial exhibitions	-	5	5
Number of seminars on development of the industrial sector	-	21	21

Source: Ministry of Information, 2006. *Sustainable Development in the Sectors of Border Areas, Communication, Industry, Mining and Energy*, Ministry of Information, Yangon:37.

existing in the country, only 8,463 are believed to be located within the prescribed industrial zones (Ministry of Information 2006:32).[2] Moreover, most of these enterprises are small in scale; many could be classified as cottage industries or family businesses. Therefore, with the exception of the garment sector, they do not generate many working opportunities for the unemployed. When US sanctions were imposed against Burma in mid 2003, government and foreign-owned factories had to close temporarily. As a result, the US State Department estimated in 2004 that of 350,000 workers in the garment sector, 40,000 to 60,000 (especially women) had been laid off (Table 8.1).[3]

A new concept: economic and trade zones on the borders

Initially, 18 industrial zones were established officially by the Myanmar authorities. Five more were added to the original list, the latest in 2006 (see Table 8.2). The 23 industrial zones are generally close to the main urban agglomerations (Ministry of Information 2006:35). The last, Thilawa-Kyauktan, is situated along the eastern bank of the Rangoon River, south of the former capital. In the near future, we can guess that another zone to accommodate the small and medium-sized non-polluting private industries will be set up in Pyinmana, close to the new capital, Naypyitaw, where sugar cane and wood-processing industries have long been established. Other recent decisions include the setting up of industrial zones in Hpa-An (the capital of Karen State, where there is a large cement factory), Moulmein-Mawlamyine (the capital of Mon State, connected with Yangon and Mandalay since February 2005, thanks to the railway bridge crossing the estuary of the Salween River) and Myawaddy,[4] announced in October 2005.

The creation of these three zones, however, reflects a new concept: the sharing of the profits derived from specific economic and industrial zones between Burma and Thailand. The proposed zones are not only industrial, they are conceived as trading centres. This concept represents the implementation of the Economic Cooperation Strategy (ECS) agreed on in November 2003 at Pagan. The participants in this program are Burma, Cambodia, Laos and Thailand, but China was also present at the meeting.[5] Vietnam joined in 2004 and the second summit of the group took place in Bangkok in November 2005. Encouraged by the success of its cooperation with Thailand, Burma is considering the creation of other economic and industrial zones on its borders with Bangladesh, India (Tamu-Moreh), China (Muse-Shweli) and Thailand (Tachilek-Mae Sai, Mae Hong Son and Kawthaung-Ranong).

The ECS provides for cooperation in five strategic areas covering agriculture, industry, trade and investment, transport, and tourism and human resources development. According to press reports of the Pagan meeting, 'the five-country economic cooperation is aimed at fully harness[ing] their huge economic potential to promote spontaneous and sustainable economic development by transforming the border areas of these countries into zones of durable peace and stability as well as economic growth'.[9] Obviously, the border zones are of prime importance for the partners, especially Burma and Thailand, as they share a 1,800km-long border.[10]

Since the Union of Burma's independence, the government—whether civilian or military—has had difficulties controlling its eastern borders due to the presence of dozens of rebel movements hiding in the deep forests. Communists, republicans, nationalists and bandits carved their petty kingdoms in these mountainous, remote parts of the country. Each group controlled one or several passages giving access to Thailand. In exchange for their protection, merchants were paying some taxes to rebels (Shan, Lahu, Pa-O, Kayah, Karen, Mon). For decades, this revenue provided the opportunity to the rebels to finance their guerrillas acting against the central government. Thai smugglers, who were dealing with all the rebels, were also taking their share from these lucrative rebellions. After 1988, with the help of China, the junta revamped and expanded several-fold the size of the *Tatmadaw* (the military, now said to number about 350–400,000 men).[11] Consequently, within a few years, most of the strongholds of the Shan and Karen rebels were retaken by the junta, which was nearly in full control of its border with Thailand. Thai smugglers, now converted into legitimate businessmen and investors, can deal openly with the

---

Table 8.2   Industrial zones in Burma, 2006

	Products
Yangon Division	
West Yangon: Hlaingthaya (453 hectares)	
Northwest Yangon: Shwepyitha	
Northeast Yangon: Shwepaukkan	
East Yangon (Mingaladon-Pyinmabin)	Textiles, food processing
Southeast Yangon: Dagon Myothit	
Hmawby-Myaungdaka	
40km north of Rangoon, 405 hectares	Steel mill, heavy industry production, plastic factory
Thanlyin (Syriam)-Kyauktan-Thilawa port[6]	
Central Burma	
Mandalay south	Railway repairs; machinery equipment, diesel engines— using imported Chinese technology—to produce small, 18-horsepower single-piston engines; soft drink food processing, soap factories
Monywa	Mechanics, gear boxes, textiles
Kyaukse	Quarries, cement plant, brick factory, bicycle factory, shoes
Myingyan	Textiles
Pakkoku	Cigarettes, textiles, mechanics
Meikthila	
Yenangyaung-Chauk	
Shan State	
Taunggyi (capital of Shan State)	Mechanics
Arakan State	
Sittwe (capital of Rakhine State)[7]	
Irrawaddy Delta	
Pyay	
Hinthada	
Myaungmya	
Pathein	
Pegu/Bago	
Burma–Thailand border zone and Tenasserim Coast	
Moulmein/Mawlamyine (400,000 rais[8])	
Mergui/Myeik/Myaik	Fisheries processing

local Myanmar authorities provided they have the proper political and economic connections. Thus transactions within the economic and trade border zone between partners who have known each other for a long time are greatly facilitated.

Meanwhile, since 2004, Burma and Thailand have developed promising bilateral cooperation in planning the creation of the first three special economic and industrial zones in Hpa-An, Mawlamyine and Myawaddy. Both countries are expecting to benefit from the establishment of these industrial zones—economically and socially. Map 8.2 shows that these areas are situated not only in the frontier space, or close to it, but on the main lines of communications connecting India, Bangladesh, China, Burma and Thailand. Moreover, projected road extensions from the border zone of Myawaddy-Mae Sod and Tak in Thailand will offer direct access towards Vietnam (Danang) and the South China Sea.

Thai investors are interested in engagement in the three combined zones and, under Thai–Myanmar cooperation, Thai factories are planning to move into the delimited areas. Participants are already engaged in cooperation, especially in the domain of energy, with the construction of the Hogyit Dam on the Salween River, in which China is also a financial partner. Construction was set to start in January 2007. As the dam is only 60km from the Thai border, factories are assured of access to a cheaper and permanent source of energy.[12]

For the Burmese side, economic and industrial zones have a triple advantage: they provide jobs to national workers; the industrial sector gets access to new technologies; and the taxes collected from the factories and traders replenish the public treasury. For the Thai side, there is also a triple advantage: factories located or relocated within the Burmese industrial zones enjoy profitable conditions (land lease for at least 75 years, profit taxes will be relaxed for re-investment with the profit earned annually); easy access to new markets (Burma, India, China) by road or by sea through the harbours of Mawlamyine or Rangoon; and especially the possibility of employing a low-paid, skilled and obedient

Map 8.2   Industrial zones along the Thailand-Burma border

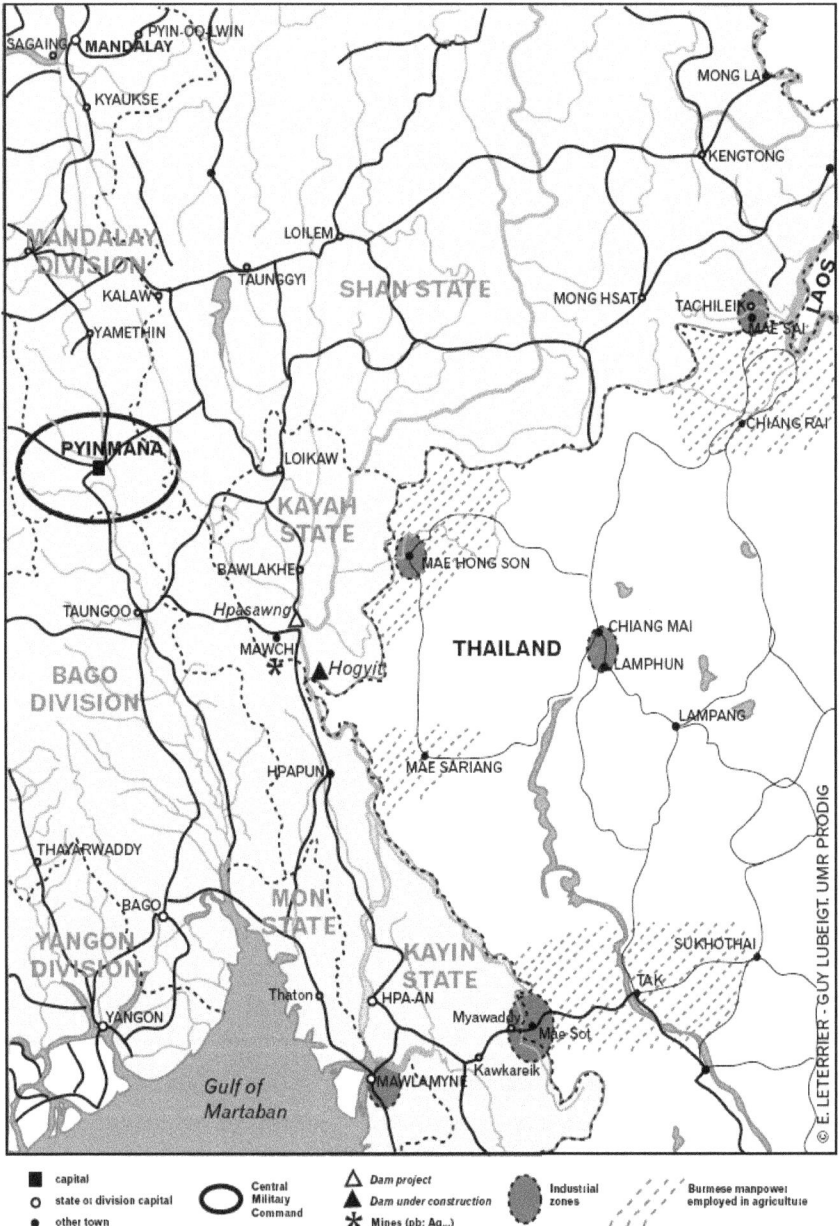

SAGAING MANDALAY PYIN-OO-LWIN

KYAUKSE

MONG LA

KENGTONG

MANDALAY DIVISION

LOILEM

TAUNGGYI

KALAW

SHAN STATE

MONG HSAT

TACHILEIK

MAE SAI

YAMETHIN

LAOS

PYINMANA

LOIKAW

CHIANG RAI

KAYAH STATE

BAWLAKHE

MAE HONG SON

TAUNGOO

Hpasawng

CHIANG MAI

LAMPHUN

MAWCHI

Hogyi

THAILAND

BAGO DIVISION

LAMPANG

HPAPUN

MAE SARIANG

THAYARWADDY

PRODIG

BAGO

MON STATE

YANGON DIVISION

KAYIN STATE

Thaton

HPA-AN

Myawaddy

Mae Sot

SUKHOTHAI

TAK

GUY LUBEIGT. UMR

YANGON

MAWLAMYNE

Kawkareik

Gulf of Martaban

© E. LETERRIER -

■ capital	△ Dam project	● Industrial zones	⫽ Burmese manpower employed in agriculture
○ state or division capital	▲ Dam under construction		
● other town	⬯ Central Military Command	✳ Mines (pb; Ag...)	

workforce. In such a win-win situation, the prospects for development of the economic and trade zones look rather bright, at least for the two governments.

For the time being, these three special zones are still in the making as negotiations between the two partners are not finalised. The questions of taxation and repatriation of the profits are not settled and potential factories are not yet in the position to create many jobs. Consequently, Burmese manpower, which cannot be employed at home, is still crossing the border in huge numbers, often paying soldiers guarding the checkpoints along the access roads to Thailand.[13]

Unemployed workforce migrates to Thailand

According to the Myanmar National Committee for Women's Affairs, using official statistics prepared by the government, the unemployment rate was 4.08 per cent in 1999 in a population of 49 million.[14] In 2001, it appeared that 4.1 per cent of the nearly 52 million inhabitants were still unemployed (Ministry of Labour 2003:8). Given the increase of the population between 2000 and 2005 (more than five million people), it appears that the employment situation not only did not improve, it was further aggravated. Other sources are no more optimistic and estimate a rise in the unemployment rate to 5.6 per cent in 2005 (CIA 2005:10). Meanwhile, anonymous local sources claim that the real rate of unemployment could be more than 20 per cent, and still rising, if we consider the number of people looking for work. The discrepancy between these sources could be explained by the fact that the authorities do not record, or pay much attention to, the unemployed who have already left Burma to find a job elsewhere. In any case, unemployment appears to be a permanent feature of the potential Burmese workforce (Table 8.3).

Official estimates from the Labour Department show that the rate of unemployment was decreasing from 4.10 per cent in 1997 to 4.01 per cent in 2002 (Ministry of Labour 2003:8), contrary to all observations. The estimated rate of participation in the workforce was then nearly 64 per cent (Table 8.3).

Table 8.3 Labour force and unemployment rate, 1996-2001

Indicator		1996–97	1997–98	1998–99	1999–2000	2000–01	2001–02
Total labour force	M	13.57	13.92	14.28	14.65	15.02	15.41
	F	8.38	8.60	8.82	9.05	9.28	9.52
	T	21.95	22.52	23.10	23.70	24.30	24.93
Labour force	M	78.57	79.09	78.63	79.19	79.68	80.26
Participation rate	F	47.18	47.64	46.67	47.01	47.32	47.65
	T	62.66	63.17	62.35	62.78	63.18	63.63
Unemployment rate	M	3.68	3.66	3.64	3.62	3.60	3.57
	F	4.77	4.77	4.76	4.75	4.74	4.73
	T	4.10	4.08	4.07	4.05	4.03	4.01

Source: Myanmar Labour Force Survey, Department of Labour (based on estimations), 2000.

If we apply that percentage in 2005, when the estimated total labour force was about 27 million workers, or half of the population, we obtain an estimation of 17.28 million workers participating effectively in the labour force. When compared with the estimated number of Burmese workers in Thailand in 2005 (more than two million), we discover that the Burmese workers in Thailand represent more than 11 per cent of the total Burmese workforce.[15]

For more than two decades, unemployed Burmese have developed informal strategies to find ways to meet their needs and feed their families. Many have illegally quit Burma and joined the foreign workforce employed in neighbouring countries. Especially attractive to Burmese workers are India, Malaysia and, in particular, Thailand. Since the beginning of the 1990s, an unprecedented flood of Burmese workers migrated to Thailand in search of job opportunities. Most of them headed towards two regions: the mountainous border space of Myawaddy-Mae Sot and, further to the south, the Bangkok area, especially the zone of Mahachai-Samut Prakan. Two other zones are known to have an important concentration of Burmese migrant workers: in the north, the border space of Tachilek-Mae Sai and, to the south, the area of Kawthaung-Ranong.

Table 8.4   Illegal migrant labour from Burma in Thailand, 1991–2004

	Arrested migrant workers from Burma	Arrested migrant workers in Thailand	Illegal migrant workers from Burma	Migrant workers in Thailand	Legal Migrant workers in Thailand	Illegal migrant workers in Thailand
1991	8,397	4,093	10,000			n.a.
1992	7,426	500,601	n.a.			n.a.
1993	n.a.	400,426	n.a.			n.a.
1994	n.a.	283,500	283,500			520,000
1995	n.a.	500,000	300,000			590,000
1996	7,664	500,134	500,000			717,689
1997	20,000	600,000	600,000			900,000
1998	290,000	350,000	1,000,000			1,300,000
1999	64,739	330,000ᵉ	600,000ᵉ			800,000
2000	130,000	330,000ᵉ	1,500,000			2,000,000
2001				568,000		
2004				1,200,000	850,000	

ᵉ estimate

Note: From one year to another registered workers can become illegal if they do not renew their registration with the Thai authorities. Many choose to do so.

Sources: *Migrations from Burma. Report 2000.* Federation of Trade Unions, 2002. 1991-93 *Bangkok Post* 13 May 1994, 28 July 1996; 1994 *Bangkok Post,* 22 January 1995; 1995 *The Nation,* January 1995; 1996 *The Nation,* July 1996 / *Bangkok Post,* 28 July 1996; 1997 *The Nation,* 28 April 1997/ *Bangkok Post, 11 July 1998;* 1998 *Bangkok Post,* 20 May 1998, 15 July 1999; 1999 *Bangkok Post,* 6 Nov 1999, 2000 *Bangkok Post,* 21 June 2000 / *The Nation,* 18 August 2000; 2001 Ministry of Labour of Thailand, 2001, 2004 Ministry of Labour and Social Welfare, Thailand.

The number of Burmese immigrants working in Thailand was estimated at between 1.5 and two million in 2000.[16] As many of them have been living illegally in Thailand for years, there is no reason to believe that their numbers would have decreased since then. On the contrary, border crossings have increased and the number of Burmese workers in Thailand, whether legal or illegal, can be safely estimated to be more than two million. It should be noted that estimations made in 2000 remain the same in 2006. Thus, it would be a surprise if the number of Burmese workers involved in the Thai apparatus of production remained the same when all information collected shows that workers rarely go back to Burma where they have no employment prospects.

Burmese women, previously estimated to represent only 20 per cent of migrant workers, are now entering Thailand in greater numbers in search of job opportunities. In Bangkok, they are in great demand in the services sector. A study conducted in 2005 by a team of researchers from Mahidol University in Bangkok revealed that about 100,000 Burmese women had taken up jobs as maids.[17] Moreover, the number of Burmese sex workers in Thailand is said to be more than 20,000. In 2000, the Federation of Trade Unions Burma estimated that more than 80,000 women and children had been sold into Thailand's sex trade since 1990. Here, too, movements have accelerated and cases of trafficking of Burmese women—sold as wives to Chinese farmers—have begun to surface. The 155,416 refugees (mostly ethnic Karen) in 11 camps along the Thai–Myanmar border, who are not supposed to work (some do), are generally excluded from estimations (Table 8.4) (Macan-Markar 2006).

A mass migration movement of such magnitude can be described only as an exodus of the Burmese workforce. The reasons inciting so many Burmese citizens to find their way to Thailand are well known and have been described at length by the media and non-governmental organisations (NGOs).[18]

Burmese workers in Thailand: registered and illegal

Unemployed Burmese travel mainly to the zones where large numbers and cheap labour are needed. Depending on locations, their number was estimated at different periods: in Mae Sot, 100,000, of whom 50 per cent were illegal, in 2006;[19] in Tak, 71,000 in 2001,[20] 200,000 in 2004; in Chiang Rai, Chiang Mai, Bangkok and Samut Sakhorn, 100,000 in 2001; in Samut Prakan and Ranong, 43,700 in 2001,[21] 100,000 in 2004;[22] in Takuapa, 10,000.[23]

Realising the scope of Burmese immigration taking place in Thailand since the end of the 1990s, the government of Prime Minister Thaksin Shinawatra signed a memorandum of agreement with the Myanmar junta in June 2003 to deport 400 Burmese nationals a month directly into a holding centre operated by the Myanmar military intelligence organisation of General Khin Nyunt.[24] Other immigrants, who were initially under the protection of the United Nations High Commissioner for Refugees (UNHCR), had to go underground to escape deportation. By the end of 2003, the Shinawatra government decided to abandon Thailand's long-standing humanitarian stance towards Burmese refugees. On 1 January 2004, the Thai authorities pressured the UNHCR to suspend its screening of new asylum-seekers from Burma. In the next months, all refugees settled in urban areas were moved and confined in camps along the border. The number of registered Burmese workers who had received temporary (one year) legal status fell from 500,000 in 2001 to 110,000 in 2003.[25]

By July 2004, the authorities reinforced their national campaign of registration of foreign workers. Attracted by an amnesty offer, 1.3 million illegal workers from Burma, Laos and Cambodia were recorded with the Thai Labour Ministry, with Burmese numbering 850,000. The administrative mechanism of registration was supposed to cope with the influx of migrant workers. Each of them had to pay 3,800 baht for a package including a medical check up, health insurance, work permit and relevant legal status. The registration was valid for one year so those who failed to re-register would become illegal immigrants again. If caught, they faced a jail sentence of up to three years and a 60,000-baht fine. Officially, no worker without a work permit would be allowed to remain in Thailand.

The registration system did not, however, put an end to the issue. Only 814,000 workers (64 per cent) were eventually issued proper work permits. Therefore, the legal foreign workers appeared to be composed of about 600,000 Burmese, 100,000 Laotians and 100,000 Cambodians. Added to that number were 93,000 migrant children less than 15 years old, among whom 80,000 were receiving no education. Most of them, illegally employed on agricultural works, could be considered victims of human trafficking.[26] Meanwhile, many migrant workers have no money when they arrive and their prospective employers are unwilling to pay the registration fee on their behalf, or to lend money to them. Moreover, the local immigration and police officials were not excessively cooperative with the scheme because the legalisation of the Burmese workers—easy prey as they were—was cutting into the lucrative possibilities for extortion. Living in constant fear of deportation, many *bona fide* migrant workers obviously escaped underground.[27] Undocumented, they unfortunately routinely experience abuse and ill treatment from employers, authorities and local communities and are threatened with arrest and deportation.

Burmese workers in Thailand

Myawaddy-Mae Sot: an industrial zone on the border

The Mae Sot area is of special interest because, in contrast with other well-known cities such as Bangkok and Phuket, this zone is situated close to the Burma–Thailand border, just opposite the old Burmese town of Myawaddy. The Burmese workers in search of employment can therefore cross the border easily to find jobs on the other side. They have only to cross a bridge over the Mae Nam Moi River (a tributary of the Salween River) to reach Thailand. The other reason why Burmese jobless choose this border crossing is that it is easily (and safely) accessible by road. Most of the workers are either urbanites or from populations living traditionally in the border area. Those coming from the hinterland have often been living in Rangoon for a few years. Thai factories (mostly textiles) relying on Burmese labour are established in and around the town and up to Tak. They can be compared with the system of the *machiladoras* on the border between Mexico and the United States.

Map 8.3   Asian highways across Burma

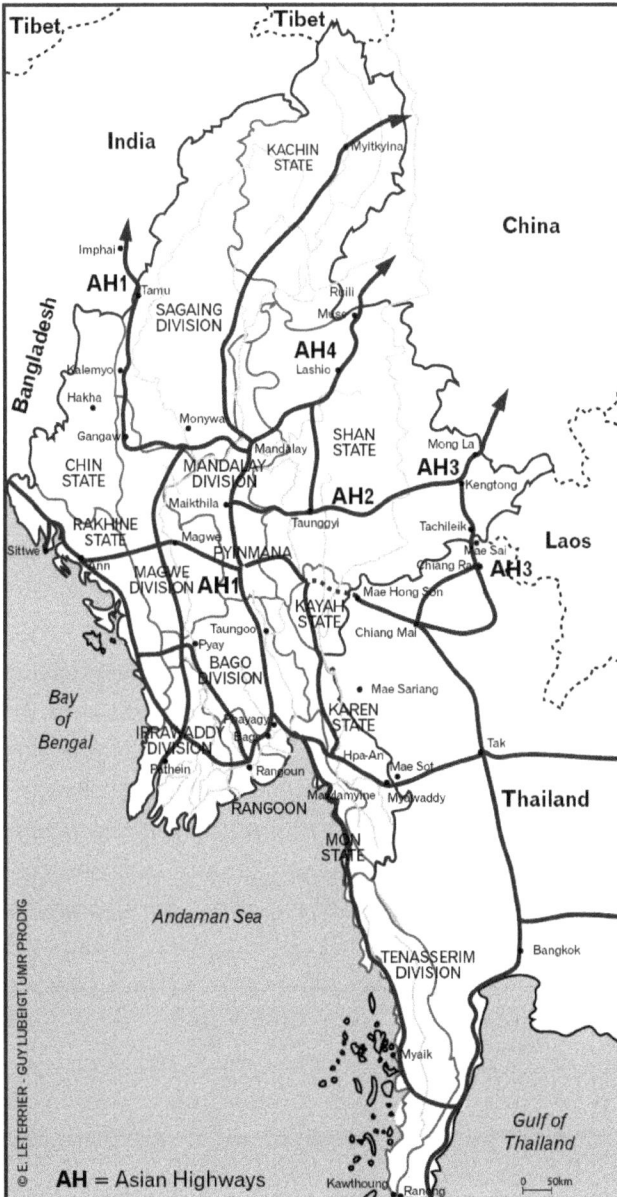

Tibet

Tibet

India

KACHIN
STATE

Myitkyina

China

Imphal

AH1

Tamu

Ruili

SAGAING
DIVISION

Muse

AH4

Bangladesh

Lashio

Kalemyo

Hakha

Monywa

SHAN
STATE

Mong La

Gangaw

Mandalay

MANDALAY
DIVISION

AH3

Kengtong

CHIN
STATE

Maikthila

Taunggyi

AH2

RAKHINE
STATE

Magwe

Tachileik

Laos

Sittwe

Ann

PYINMANA

Mae Sai

MAGWE
DIVISION AH1

Chiang Rai

AH3

Taungoo

KAYAH
STATE

Mae Hong Son

Pyay

Chiang Mai

BAGO
DIVISION

Bay
of
Bengal

Mae Sariang

Bago

KAREN
STATE

IRRAWADDY
DIVISION

Nyagyi

Hpa-An

Mae Sot

Tak

Pathein

Rangoun

RANGOON

Mawlamyine

Myawaddy

Thailand

MON
STATE

Andaman Sea

Bangkok

TENASSERIM
DIVISION

© E. LETERRIER - GUY LUBEIGT UMR PRODIG

Myaik

Gulf of
Thailand

AH = Asian Highways

Kawthoung

Ranong

0   50km

That road is part of the Asian Highway One (or, some say, 'the old Paris–Saigon road'), which, from India and the border town of Tamu, enters Thailand at Mae Sot, runs down to Bangkok and then across to Ho Chi Minh City through Phnom Penh. Now Asian Highway One also links China with its branches, AH2, AH3 and AH4 crossing in Mongla, Muse and Kachin State. Soon it will also be connected to Bangladesh. Asian Highway One is part of the East–West economic corridor that will link the transportation networks of central Vietnam and Burma through Laos and Thailand as part of the Greater Mekong Subregion development program.

The Myawaddy-Mae Sot border area has, however, a major inconvenience for the jobless: it is still partly a war zone, with fighting between the *Tatmadaw* and the leftover Karen and Shan rebel forces (KNU and Shan State Army South). Therefore, some migrants proceed south on the Mawlamyine–Tavoy road (which is also sometimes cut by the operations of small groups of Karen and Mon rebels) to reach Mergui. From there, they can cross the Tenasserim Range in the valley following the Tenasserim River up to the Maungdaung Pass, which has been used since prehistoric times to reach the Gulf of Siam. Most migrants, however, prefer to go to Kawthaung from where they can cross easily to the border town of Ranong. From Ranong, job-seekers can be recruited illegally by many Thai agents who dispatch them (after payment) either further south (to Phuket) or towards Samut Prakarn-Bangkok, where there is a big demand for Burmese manpower.

The Myawaddy-Mae Sot zone, long an attractive area for ordinary Burmese workers, is now also attractive for the Burmese and Thai authorities—so much so that the border area, already chosen for development as a common Special Economic and Industrial Zone between the two countries, is going to become the second largest such zone in Burma.[28] This pilot project of cooperation, much inspired by the Chinese economic zones, is founded on the existence of available manpower, which could be employed in the factories planned for Burma, on the other side of the border. More economic and trade border zones will follow, including in Tamu (the border town with

India) and Maungdaw (the border town with Bangladesh). As Burma has 13 main border-trade points with its four neighbouring countries, the prospects are large.

Joint exploitation of human resources

Expanding its cooperation with Thailand, Burma has recently agreed to lease fallow lands to Thailand in the border areas (Tachilek, Mae Hong Son, Mae Sariang, Hpa-An, Ye).[29] Under the system of contract farming, seven million hectares have been reserved for cultivation. Burma agreed to plant crops (sugar cane, oil palm, cassava and rubber) that will be exported and processed in the economic and industrial zones or in Thailand. This system is conducted under the Irrawaddy–Chao Phraya–Mekong Economic Cooperation Strategy. The practice is applied to agricultural crops that are labour intensive and is aimed at import substitution. The system is intended to generate jobs in neighbouring countries and will support investment expansion between partners. Further, it is hoped to effectively solve illegal border crossings by migrant workers, reduce health and social problems originating from illegal labour and patch up the differences in development levels between Thailand and its neighbours, especially Burma. Basically, Thai goods will become cheaper because the raw materials and the labour will be cheaper in the other countries.[30] Whatever the reason given, Burmese and Thai authorities have jointly decided to exploit fully the human resources available in Burma. The Burmese side wants to get a share of the profits generated by the exploitation of its workforce in Thailand,[31] while the Thai side will gladly take advantage of the cheap manpower offered by the Burmese junta.

Burmese workers (because they are usually illegal) are passive, easy to handle and easy to bully; no strikes and no demands for the amelioration of working conditions are likely from them. They are often not provided with housing, running water or medical care, if they are not registered, and their working hours are extended at will (often nine to 16 hours a day, poorly paid, or with no pay at all). Any rebellion is punished immediately by expelling the recalcitrant workers to the other side of the border; or by threatening or beating them. It can go as far as assassinations.

Through a network of seasoned smugglers and traffickers well connected with the Burmese army on one side, and with police and immigration officials on both sides, Burmese labourers are brought to their destinations: the factories in Thailand. Thai and Burmese brokers, police and immigration officers connected with factories owners and human traffickers organise the recruitment—and sometimes the sale—of workers.

Migrant workers: the new slaves of the globalised world

The influx of migrant workers is so huge that the employment possibilities have spilled over into the Mae Sot economic zone towards Tak. Burmese workers are in great demand everywhere. They are employed not only in factories, but in agriculture (orchards), construction, as housemaids and, of course, for prostitution. Among many advantages, the Burmese worker can be easily dismissed or replaced, because, when a position is vacant, several new workers (often waiting on the other side of the border) are ready to come in and take a job for any salary at all. Above all, Burmese workers are cheap. Desperate in Burma, they accept between 40 and 60 baht a day, which is only one-third of what would be paid to Thai workers.[32] Compared with salaries paid in Burma, even after the government salary increases of 1 May 2006 (up to US$30 a month for a high-level government civil servant previously receiving US$5–6 a month), this is still quite a boon for the jobless (Table 8.5).

Burmese migrant manpower in Thailand is exploited at will by Thai employers. Surveys suggest that these factories are often owned by ethnic Chinese (from Thailand, Taiwan, Hong Kong, Malaysia or Singapore). For these entrepreneurs, the only goal is profit by any means. And, for them, Burmese workers—as foreigners with no rights to stay or work in Thailand—are no more than expendable spare parts. Like many migrant workers throughout the world, the Burmese workforce can therefore be included in a new category of workers: the slaves of the globalised world.[33]

Table 8.5   Types of jobs and salaries paid to Burmese workers in Thailand

| No. | Types of jobs | Wages/Salaries | | | |
| | | Daily | | Monthly | |
		baht	US$	baht	US$
1	Agriculture			2,500	71.42
2	Cold storage			2,500	71.42
3	Construction	100–180	2.29–4.57		
4	Coastal fishing			2,000	57.14
5	Fishery (deep-sea)			2,500	71.42
6	Food shops			2,000	57.14
7	Gas stations			1,500	
8	Gem cutting			3,500	
9	Gem mining			2,000	57.14
10	Golf courses			2,000	57.14
11	Industry	150–180	2.85–4.57		
12	Housekeeping*			1,000	28.57
13	Logging	80–150	2.28–4.28		
14	Saw mills	80–160	2.28–4.57		
15	Services	90–160	2.57–4.57		
16	Restaurants*			2,000	57.14
17	Rice mills			2,500	71.42
18	Rock grinding	80–160	2.28–4.57		
19	Slaughter houses			2,000	57.14
20	Paper mill	90–160	2.57–4.57		

*with a place to stay. Maids work nine to 16 hours a day.
Note: The situation remains the same in 2006.
Source: Federation of Trade Unions Burma, 2002. *Migrant Workers from Burma*.

Amazingly, at the same time, when Sino–Thai employers exploit their Burmese, Laotian and Cambodian 'slave workers' in Thailand,[34] Thai workers employed abroad (in Taiwan) have to revolt against the working conditions imposed on them by their Taiwanese employers in Taipei. These conditions, described as inhumane by the Thai workers themselves, led in August 2005 to riots against the Chinese employer. Thai press reports of the events, curiously, did not draw a parallel between the situation of the Thai workers in Taiwan and that of the Burmese workers in Thailand.[35]

In Singapore, foreign workers are mostly legal and protected by law when working in factories. Individual employers, however, seldom respect the law and also exploit their Burmese maids.[36] The Singaporean press mentions and sometimes denounces the actions of Chinese employers (especially abuses against maids) when such cases happen to be brought to court. But this occurs only rarely in Thailand, and then it is principally because Thai government representatives in foreign countries immediately take up the abuses of the employers in order to help or protect their fellow citizens, whether the latter have migrated legally or not.[37] In Burma, the government-controlled press never mentions the plight of the expatriate Burmese workers in Thailand, mainly because they often left Burma illegally.[38] When Burmese workers have proper documentation, the Burmese authorities gladly accept their monthly remittances in Burma and charge a fee to renew their passport in Thailand every two years (10 per cent of their salary), but they do not bother to intervene on their behalf. If a worker is in conflict with his or her employer, the reaction of the Burmese representation is always the same: the Burmese worker involved is immediately sent back to Burma, and the matter is then considered closed.

In the border zones, where no Burmese consulate exists, the workers are abandoned to their fate and receive no protection from their authorities whatsoever. Whether in Thailand, Malaysia or Singapore, local laws are not made for the benefit of migrant workers. On the contrary, the tendency of the authorities is to use the migrant workers as pawns for bargaining.

A Thai economy fed by cheap manpower

For more than a decade, the Thai Immigration Police have regularly announced new measures to curb immigration—legal and illegal—from neighbouring countries. Thai policy fluctuates between crack-downs and registration, arrests and bans, or forced repatriation and expulsion, but to little avail. Burmese job-seekers (as well as Laotian and Khmer) can sneak into Thailand. While the Thai government is at pains to control its porous borders,[39] the foreign workforce is secretly encouraged to come in to work for employers willing to take advantage of the cheap cost of their labour.

Meanwhile, since 2006, Thai entrepreneurs and investors have realised that a serious problem is looming for the Thai economy. There will soon be severe labour shortages in seven key industries (petrochemical, food, automotive, tourism, textiles, software and logistics) because Thai schools do not produce the right personnel to serve the expansion of the economy. The Education Ministry recently announced that, within three to five years, there will be a shortfall of almost 585,000 workers in Thailand.[40] Moreover, some managers already complain that it is difficult to employ Thai workers coming directly from farms to the factories. These workers have to follow a special formation to enter fully the line of production. In such cases, the local authorities (and the government) are always prompt to react in favour of the investors and cater to their manpower needs. Unfortunately, the education sector has been neglected by the government since 2001 and promised reforms have not taken place.[41]

After a decade of tough policy designed to return illegal immigrants to Burma,[42] a new policy was set up on 1 October 2006 to allow in a special quota of skilled workers requested by seafood-processing factories and fishing trawlers—areas in which Thais do not want to work. Given that the evaluation of the need was clear, the solution chosen by the Thai authorities was rapid. While the government was monthly expelling 10,000 Burmese migrants from Thailand in 2003, under the program of 'informal deportation', it will now facilitate the official entry of skilled Burmese workers into Thailand to take up the positions available in its factories. A shift in Thai policy towards migrant workers began in May 2005 with authorisation given to foreign workers residing in border areas to come to work in Thailand providing they returned to their homes every evening. The hinterland factories relying on the cheap Burmese labour are, however, hungry for manpower. Entry permits for Thailand recently delivered to 10,000 skilled Burmese workers are only the beginning of a new process; more will follow to cater to the needs of the Thai economy.

In continental Southeast Asia, workers fuelling the economy are recruited preferably from neighbouring countries with a manpower surplus.[43] According to Chidchai Vanasatidya,[44] 'there will be no illegal immigrants living or working in Thailand by the end of next year.' A new mechanism to control the foreign workers will then be fully operational. The Immigration Police will lead 10 other government agencies responsible for national security and civilian intelligence services in stamping out illegal entry, unlawful employment and human trafficking. Some 100 million baht will be spent on creating a database containing details of all foreigners entering, leaving, living and working in Thailand. The system will certainly work with all the *bona fide* travellers presenting their passports at the different ports of entry into Thailand, but its efficiency is doubtful when confronted with the illegal Burmese, Laotian and Cambodian migrants willing to enter the country to find work, or with the illegal Chinese determined to use Thailand as a departure base for economic migration to Western countries.

Conclusion

Massive migration of Burmese workers into Thailand affects both countries. On one hand, it depletes the availability of skilled workers in Burma, which is a clear loss for a developing country, while on the other hand, Thailand benefits from such a reservoir of cheap manpower. Burma receives the monthly remittances of its expatriate workers, but Thai entrepreneurs capitalise on the value added to their export-oriented productions by the work of the Burmese migrants.

Each country is aware of the size of the phenomenon and its impact on their economy, but each reacts differently. The Myanmar junta chooses to ignore the huge emigration taking place, because it reduces the potential of social, if not political, demands building up within society. The Thai government plays down the boost given to its economy by the widespread use of cheap Burmese workers by its industries, and prefers to play up the supposed or real social disorders said to be brought by Burmese immigrants: increase of diseases such as malaria, tuberculosis and HIV; the drain on hospital resources to care for sick Burmese;[45] the expansion of prostitution; and murders and thefts. The dual attitude of the Thai authorities is politically useful to hide their own social and health shortcomings from their own population. The contribution of migrants to the Thai economy is still unrecognised officially, although a 'new vision' towards migrants is beginning to appear in government circles, probably out of necessity and to be in accordance with the Economic Cooperation Strategy illustrated by the launch of the first economic and industrial zone in Myawaddy-Mae Sot. For their part, Burmese authorities, until now ignoring the plight of their expatriate workers, recently realised the potential political benefits of monitoring such a huge workforce in Thailand.

Notes

1   Inhabitants of these townships need one to two hours of travel on dilapidated buses to reach the downtown.
2   Other statistics record more than 100,000 'industries' employing more than two million workers in the whole of Burma.
3   In 2002, Burmese exports to the United States, mostly garments, were worth US$356 million. Garments were then the second largest export. By September 2005, the Burmese garment industry revived with the imposition of US and EU quotas on Chinese imports. American and European retailers immediately boosted their orders from other low-priced suppliers. Since then, private sources explain that big factories (often South Korean-owned) have quietly reopened in Burma and export their production to South Korea, where labels are stitched on before being re-exported to the United States and Latin America as South Korean products.
4   Covering 396, 275 and 384 hectares respectively. The Myawaddy zone is a pilot project that will be the prolongation, in Burmese territory, of the Mae Sot industrial zone.
5   This project is also known as the Ayeyawaddy (Irrawaddy)–Chao Phraya–Mekong Economic Cooperation Strategy (ACMECS). Three participants (Burma, Laos and Vietnam) share a common border with China, but all share the waters of the Mekong River, which are controlled increasingly by China.
6   Established in May 2006 as a Special Industrial Zone with 100 per cent foreign investment, the Thilawa port zone will be the first export-concentration zone in

which all formalities for export of the zone's products will be handled. Most, if not all, of the investments for the construction have been provided by China. Therefore the law governing the zone includes restrictions on investment by domestic national entrepreneurs unless it is done with joint investment from foreign counterparts.

7   Development project with a special industrial zone based on natural gas from offshore.

8   Many plots located between Hpa-An (capital of Karen State) and Moulmein (capital of Mon State) have been allotted to be parts of an economic and industrial estate to be built within the framework of Burma–Thailand cooperation. The zone will be connected with the industrial zone of Mae Sot. The *rai* is the Thai unit for measurement of area (1 rai = 1600 square metres).

9   *Xinhuanet* reporting from Yangon, 30 October 2005.

10 Burma and China also share 2,185 kilometres of common border along the Yunnan Province.

11 For an assessment of the size of the army, see also Mary Callahan's chapter in this volume.

12 The Hogyit Dam is also located almost equidistant between Naypyitaw and the industrial zones of Hpa-An, Mawlamyine and Mae Sot. It is the first of the five gigantic hydroelectric plants planned on the lower course of the Salween River by Burma and the Electricity General Authority of Thailand. The dams will have a combined capacity of 11,800 megawatts (MW), with the giant Ta-Sang having 7,110MW. The dams are said by Thailand to be necessary to ensure adequate energy supplies in the region. The actual capacity of Burma is 1,335MW, up from 707MW in 1988.

13 The author observed such a check-point in March 2005 on the road from Thanbyuzayat to Ye on the Tenasserim coast.

14 Myanmar National Committee for Women's Affairs 2001:28 (from the *Handbook on Human Resources Development Indicators*, 2000). This organisation, presided over by wives of generals, is fully controlled by the military regime.

15 In 2002, the number of Burmese workers in Thailand was estimated to be between 7.9 and 10.5 per cent of the total Burmese workforce. See Lubeigt 2002:1.

16 General Sarit Sarutanon, Joint Chief of the National Police, *Bangkok Post*, 21 June 2000. Cited in Caouette and Pack, 2002.The estimation is reproduced in Federation of Trade Unions Burma 2002. The Federation is an exiled organisation connected with the democracy movement.

17 *Bangkok Post*, 29 October 2005.
18 See the well-documented field reports of Battistella, G. and Asis, M.M.B., 1999; Caouette et al. 2000; Huguet 2005; Horton 2006; Human Rights Watch 2004; and BHPWT August 2006.
19 *Democratic Voice of Burma*, 5 July 2006.
20 Arnold, D., 2004
21 Arnold, D., 2004
22 According to Naing 2004, Ranong itself had 40 karaoke bars and brothels; and 150 Burmese prostitutes operated their trade in the town. Given the number of Burmese workers living in the area, this survey remains doubtful.
23 Unitarian Universalist Service Community, January 2005.
24 Shinawatra abandoned Thailand's traditional policy in order to please the junta, with which he had several business deals. For Burmese intelligence, the aim was mainly to get information on the resistance movements developing in Thailand in the wake of the attack on Nobel Prize-winner Aung San Suu Kyi (on 30 May 2003) by a mob of Union for Solidarity and Development Association (USDA) members, sponsored by the junta.
25 *Bangkok Post*, 15 July 2004.
26 *The Nation*, 24 August 2005.
27 It is difficult to differentiate between asylum-seekers, refugees, migrant workers, refugees having to work to sustain their families and political refugees or those exiled. Most of the migrants fit into several categories at the same time. Their common identity is that they have all been driven out of Burma by the policies of the military regime.
28 The 105th Mile Border Trade Zone, established on the Burma–China border of Yunnan Province in April 2006, is the first zone in terms of importance. The manpower in the zone is, as in Mongla, actually Chinese.
29 Possibly taken from 'relocated' Shan and Karen villages; 600,000 to one million members of ethnic minorities are said to have been internally displaced by the Burmese junta.
30 *Bangkok Post*, 1 December 2005.
31 Another, political benefit seems to emerge for the Burmese authorities, who want their illegal workers to come back to Burma to register. After verification of their ethnic nationality, travel documents will be issued to certified workers to return to Thailand. This new regulation results from a memorandum of understanding signed between Senior General Than Shwe and Prime Minister Shinawatra during the last unpublicised visit of the latter to Burma in August 2006. Only belatedly aware of the agreement, the Thai press denounced the move as an attempt to implant more pro-junta supporters in Thailand,

especially those related to the USDA, in order to spy on the activities of the Burmese community living there. See Chongkittavorn 2006:9.

32 The daily official minimum wage is set by the Thai government. In May 2005, it was 175 baht in Bangkok; 173 baht in Phuket; 149 baht in Chiang Mai; 147 baht in Ranong; and 139 baht in Mae Sot.

33 The situation is better for the Laotians as they speak Thai, while the Cambodians often understand the Issan dialect of northeastern Thailand.

34 The idea of slavery, now being discussed without provoking horror or widespread condemnation, seems to be accepted. See *Bangkok Post* 2005.

35 In mid December 2006, as Thai industries needed more labour, the press campaigned for the protection of the rights of Burmese workers. See Chongkittavorn 2006:9.

36 The number of Burmese maids employed in Singapore (currently paid S$300–450 monthly) could be as high as 50,000. One of them claims that 80 per cent of the Chinese bosses are 'very bad with their employees'. Maids have one day off twice a month. Another complained that after her arrival she had no day off for four years (verbal communication, 20 July 2006).

37 Mainly to avoid being criticised, the Thai press is always prompt to defend the rights of the Thai citizens unfairly treated.

38 Official documents are costly (US$3,000), difficult to obtain (taking three to six months) and there is no guarantee of finding a legitimate job. Only seamen, who are in great demand as they speak English, are officially recruited by private agencies in Burma. They leave officially and benefit from the 'protection' of the government because they send back handsome (by Burmese standards) monthly remittances. The amount sent back by the workers to Burma is said to be $6 billion ($1.5 billion for the 150,000 Thai overseas workers). *The Nation*, 24 August 2005.

39 To control its more than 5,000 kilometres of border, Thailand deploys 3,800 immigration officers. *The Nation*, 4 August 2006.

40 *The Nation*, 4 August 2006. Details from the previous registration show that 177,226 Thai employers wanted to hire 1,087,834 migrant workers. At the same time, 13,487 migrant workers entering Thailand were arrested and 7,354 of them are now being prosecuted. *The Nation*, 7 September 2006.

41 There have been six ministers of education in six years.

42 One can wonder whether such a policy was designed to effectively send back the illegal workers to their country of origin, or to keep pressure on the foreign job-seekers willing to accept wages amounting to one-third of the salaries that should legally be paid to Thai workers (see note 39).

43 According to AFP (22 June 2005), Malaysia is one of Asia's largest importers of labour. Foreign workers, legal and illegal, number about 2.6 million of its workforce of 10.5 million.

44 Then deputy Prime Minister in the Shinawatra caretaker government. He was holding the portfolio of Home Minister jointly until he was ousted by the coup of 19 September 2006. *The Nation*, 4 August 2006.

45 According to the Public Health Ministry Inspector-General, Dr Kitisak Klabdee, the Burmese patients' unpaid medical bills at hospitals in Tak were expected to exceed 70 million baht in 2005, because hundreds of thousands of them living near the border had no medical insurance. *Bangkok Post*, 29 June 2005.

References

Agence France Presse, 2005. 'Malaysia to ban abusive employers, consider letting refugees work: reports', *Agence France Presse*, 22 June.

Arnold, D., 2004. 'The situation of Burmese migrant workers in Mae Sot, Thailand', Working Paper Series, No. 71, September, Southeast Asia Research Centre, City University of Hong Kong, Hong Kong.

Backpack Health Worker Team, 2006. *Chronic Emergency: health and human rights in Eastern Burma*, September, Backpack Health Worker Team, Bangkok.

Bangkok Post, 2005. 'Burmese maids get treated like "slaves" in Thailand', *Bangkok Post*, 29 October 2005.

Bangkok Post, 2005. 'Myanmar agrees to cultivate seven million hectares under contract farming with Thailand', *Bangkok Post*, 1 December.

Battistella, G. and Asis, M.M.B., 1999. *The Crisis and Migration in Asia*, Scalabrini Migration Center, Quezon City.

——, and Skeldon, R., 1999. Towards regional cooperation on irregular/undocumented migration, International Symposium on Migration, Bangkok, 21–23 April 1999, International Organisation for Migration and Scalabrini Migration Centre, Bangkok and Manila.

Callahan, M.P., 2003. *Making Enemies. War and state building in Burma*, Cornell University Press, Ithaca.

Caouette, T. Kritaya, A. and Pyne, H. H., 2000. *Sexuality, Reproductive Health and Violence: experiences of migrants from Burma in Thailand*,

Institute for Population and Social Research, Mahidol University, Bangkok.

Caouette, T.M. and Pack, M.E., 2002. 'Pushing past definitions: migration from Burma to Thailand', Refugees International and Open Society Institute, Washington DC.

Central Intelligence Agency (CIA), 2005. *The World Factbook, Burma*, Central Intelligence Agency. Available from http://www.cia.gov/cia/publications/factbook/geos/bm.html

Chongkittavorn, K., 2006. 'Thailand's cynical ploy on Burmese migrant workers', *The Nation*, 11 December 2006:9.

Chulalongkorn University, 2003. 'Reviewing Policies and Creating Policies to Protect Migrant Workers', conference organised by the Institute of Asian Studies, Bangkok.

Evans, G, Hutton, C.M, Kuah, K.E., 2000. *Where China Meets Southeast Asia*, White Lotus, Bangkok.

Federation of Trade Unions Burma, 2002. *Migrant Workers from Burma*.

Horton, G., 2005. *Dying Alive: a legal assessment of human rights violations in Burma*, Report to the Netherlands Ministry of Development Cooperation, Images Asia, Chaingmai. Available from: http://www.ibiblio.org/obl/docs3/Horton-2005.pdf

Huguet, G.W. and Sureeporn, P., 2005. *International migration in Thailand*, report to the International Organisation for Migration, 23 August 2005, Bangkok. Available from http://www.old.iom.int/documents/publication/international_migration_thailand_23_aug_05.pdf

Human Rights Watch, 2004. *Human Rights Watch World Report*, Human Rights Watch, New York.

Kasem, S., 2005. 'Burmese to leave B70m in unpaid bills', *The Nation*, 29 June.

Lubeigt, G., 1989. 'Aspects du développement de la zone péri-urbaine de Rangoun (Birmanie)' [Notes on the Development of the peripheral urban zone of Rangoon], in *La Péri-urbanisation dans les pays tropicaux*, Bordeaux, CNRS-CEGET, *Espaces Tropicaux*, No.1:327–84.

——, 1993. 'Les dimensions socio-politiques et culturelles de la périurbanisation en Birmanie' [The social-politics and cultural dimensions of the peripheral urbanisation in Burma], in *Métropolisation et périurbanisation*, Cahiers du CREPIF, No.42 (March):165–6.

——, 1994. 'Birmanie: une nouvelle péri-urbanisation' [Burma: a new peripheral urbanisation in Burma], in *Penser la ville de demain-Qu'est-ce qui institue la ville?*, L'Harmattan, Paris:203–14.

——, 1995. 'Données stratégiques d'un aménagement urbain en Birmanie: de Rangoon à Yangon' [Strategic Facts of an Urban Development in Burma: From Rangoon to Yangon], in *Cités d'Asie*, Les Cahiers de la Recherche Architecturale, Nos 35–6, Editions Parenthèses/diff. PUF (March):141–51.

——, 1998. 'Réflexions sur l'espace frontalier birmano-siamois et ses enjeux traditionnels (XIIIème–XIX ème siècles)' [Reflections on Burma-Siam Frontier and its Traditional Stakes (thirteenth –nineteenth Centuries], in *Guerre et Paix en Asie du Sud-Est*, Collection 'Recherches Asiatiques', Sophia University (Tokyo) et L'Harmattan, Paris:9–38.

——, 1999. 'Ancient transpeninsular trade roads and rivalries over the Tenasserim coasts', in *Trade and Navigation in Southeast Asia (fourteenth–nineteenth centuries)*, Collection 'Recherches Asiatiques', Sophia University (Tokyo) et L'Harmattan, Paris:47–76.

——, 2002. Esclaves du monde moderne: les travailleurs migrants birmans en Thaïlande, Communication to the conference of Reseau Asie, CNRS, Paris, September.

Macan-Markar, M., 2006. 'Ethnic groups pin hopes on visiting UN official', *Inter Press Service*, 17 May.

Maneepong, C., 2006. 'Regional policy thinking and industrial development in Thai border towns', *Labour and Management in Development Journal*, 6(4), The Australian National University and University of Tasmania, Asia Pacific Press, Canberra. Available from http://labour-management.anu.edu.au/volumes/prt/volsix/6-4-Maneepong.pdf

Maung Aung Myoe, 2002. *Neither Friend Nor Foe: Myanmar's relations with Thailand since 1988, a view from Yangon*, Institute of Defence and Strategic Studies, Singapore.

Ministry of Information, 2006. *Sustainable Development in the Sectors of Border Areas, Communication, Industry, Mining and Energy*, Ministry of Information, Yangon.

Ministry of Labour, 2000. *Myanmar Labour Force Survey*, Department of Labour, Yangon.

Ministry of Labour, 2003. *Handbook on Human Resources Development Indicators, 2002*, UNFPA and Department of Labour, Yangon.

Myanmar National Committee of Women's Affairs, 2001. *Gender Statistics in Myanmar*, Myanmar National Committee of Women's Affairs, Yangon.

Naing, W., 2004. 'The trafficking trap persists in border towns', *Inter Press Service*, 18 December 2004.

Rajah, A., 1994. Burma: protracted social conflict and displacement, Paper presented at the International Conference on Transnational Migration in the Asia Pacific Region: Problems and Prospects, Bangkok.

Santimatanedol, A., 2005. 'Migrant workers "exploited"—study: Burmese maids get treated like "slaves"', *Bangkok Post*, 29 October.

Shakti, R.P., Supang, C. and Naing 1997. *Reproductive Health Survey: migrant Burmese women in Ranong Fishing Community, Thailand*, April, Institute of Asian Studies, Bangkok.

Stern, A. and Crissman, L.W., 1998. *Maps of International Borders Between Mainland Southeast Asian Countries and Background Information Concerning Population Movements at these Borders*, February, Asian Research Center for Migration, Institute of Asian Studies, Chulalongkorn University, Bangkok.

Taylor, R.H., Tin Maung Maung Than and Kyaw Yin Hlaing (eds), 2005. *Myanmar: beyond politics to societal imperatives*, ISEAS Publications, Singapore.

The Coalition to Stop the Use of Child Soldiers, . *Child Soldiers: country briefs*. Available from http://www.child-soldiers.org

The Nation 2006. 'Illegal immigration should end next year', *The Nation*, 4 August.

The Nation, 2006. 'Workers registered Oct 1', *The Nation*, 7 September.

Tin Maung Maung Than, 1999. Mimicking a developmental state: Myanmar's industrialization effort (1948–62), The Myanmar Two Millennia Conference, 15–17 December, Yangon.

United Nations Interagency Project on Human Trafficking, . *UN Interagency Project on Human Trafficking in the Greater Mekong Sub-Region*, United Nations Interagency Project. Available from http://www.no-trafficking. org/Publication/publications.html

Wilson, T. (ed.), 2006. *Myanmar's Long Road to National Reconciliation*, ISEAS Publications and Asia Pacific Press, Singapore and Canberra.

9   Environmental governance in the SPDC's Myanmar

Tun Myint

The speed of environmental transformation in Burma has been intensified since the State Peace and Development Council (SPDC) took power by violent military coup on 18 September 1988. Desperately needing financial capital to sustain its military power and engage in political and armed annihilation[1] of various insurgent groups (ABSDF Research and Documentation Center Office [unpublished]; Lintner 2002), the regime began indiscriminately exploiting the country's natural resources. The reality, however, is that the institutional development for environmental governance falls behind the intensified uses and abuses of the natural environment in Burma. More imprudently, the benefits generated by intensification of environmental exploitation do not improve the well-being of citizens, let alone contribute to the economic development of the country.

In 1989, the SPDC adopted an open economic policy and the announcement of its 'open-door' policies soon attracted foreign investment. The flood of foreign investment into various sectors of the economy raised concern about environmental issues. At the Earth

Summit Plus Five in 1997, the then Burmese Foreign Minister, U Ohn Gyaw, asserted that Burma's environmental problems were a result of 'underdevelopment'. As the nation strived to catch up with the rest of the world in terms of material development, natural resources became primary targets for development capital in Burma. This national quest and campaign for development posed the dilemma of 'sustainable development'.

The 18-year long political stalemate between the military regime and the democratic opposition led by the National League for Democracy (NLD) has left Burma with no constitution, no national legislative body and no independent judicial system. In other words, Burma lacks the fundamental structures of a stable society—such as political accountability, good governance and effective and equitable enforcement of the rule of law—that are essential for the sustainable management of environmental and human resources.

At this current political juncture, the lack of rule of law and good governance mechanisms poses challenges and limited opportunities for meaningful environmental governance in Burma. Taking this political condition as a fundamental basis of the challenges for environmental governance, this chapter first assesses the limited opportunities to strengthen environmental governance. Second, it discusses strategies for how people with concerns about environmental issues could overcome the challenges that exist in the current political context. Third, the chapter concludes that the primary responsibility to improve environmental governance lies in the hands of the SPDC. Three questions guide this assessment. First, how does the political instability in Burma influence issues of environmental governance? Second, what are the limited opportunities for environmental governance in the continuing political crisis in Burma? Third, how can environmental issues be utilised to create political space for self-governance and more freedom for the local population, and in turn contribute to political transition as a whole?

Before advancing further, the risks and challenges of analysing any policy in the military regime's Myanmar deserve mention. Without the accompanying normal assessment and analysis of history, and in the current state of political crisis, any analysis of environmental policy in the SPDC's Myanmar will be a mere review of the regime's policy on paper. In Burma, official policies and even laws are usually not followed by the military rulers themselves. The country is still ruled by martial law[2] at best and one-man authoritarian rule at worst. Therefore, academic study of any type of policy and governance in Burma under the current military regime involves risks and challenges that come not only from the regime's control of the form and shape of official data, but from the self-censorship and information manipulation of élitist Burmese[3] and international non-governmental organisations (INGOs), including the United Nations' mechanisms, working inside Burma.

Environmental transition and related issues

Environmental transition in Burma continues while political transition faces deadlock. Much of the literature and news reports focus on a narrow sense of political transition (power transfer) without paying attention to other areas, such as economic and environmental transition. In developing countries such as Burma, politics is mainly about control over land and natural resources. In short, environmental transition is no less important than political transition, which is conceived predominantly as the transfer of power from the military to a civilian government. This chapter draws attention to environmental transitions and issues under the SPDC's rule in Burma.

Overview

As one of the most fertile and mineral-rich countries in Asia, Burma is a land of 'stunning ecological diversity' (Smith 1994:12), which is reflected in the existence of the diverse cultures, histories and traditions of the many ethnic groups living with nature in the highland and lowland areas of the country. Ecosystems in Burma vary from tropical islands, rainforests and lush tracts of mangrove to great rice-growing plains in the south and snow-capped peaks of mountain pine in the north.

In addition, Burma is endowed with a rich diversity of habitat types arising largely from its unique ecological diversity. Two independent biodiversity assessments, the World Resources Institute's *Last Frontier Forests* and Conservation International's *Global Biodiversity Hotspots*, rank Burma among the top priority countries in mainland Southeast Asia, along with Laos and Cambodia.

This environmental endowment, however, and Burma's status as being rich in natural resources, which is often proudly claimed by the military leaders, is threatened by the almost unfettered exploitation of natural resources occurring within the current political crisis. In recent years, Burma has seen major resource-exploitation development projects proposed, and in some cases implemented, with large-scale impact on the natural environment and on natural resource endowment. The controversial Unocal/Total gas pipeline project to Thailand, secret and often illegal logging concessions, the Shwe field gas exploitation led by Korean firm Daewoo International, and the Salween Dam schemes were not approved by Parliament, were not conducted with any transparency, and were not the subject of any kind of public debate or consultation with the local population who will be directly affected.[4] There were no meaningful social and environmental impact assessments for these projects. Considering the current political context in which the SPDC single-handedly conducts these development projects, without the proper participation of the local population and domestic stakeholders, one can assert that the environmental endowment of Burma is being handed over to SPDC-owned enterprises and foreign investors. In this imprudent state of environmental exploitation, the main beneficiaries are not the Burmese people or the Burmese army. Although this chapter does not address who the beneficiaries are, it is worth asking about the nature and origins of these beneficiaries because the SPDC regime would not have survived without them injecting funds to maintain its existence.

Agriculture

Burma's economy is dominated by the agricultural sector, which generates more than 50 per cent of total GDP and employs more than

60 per cent of the total labour force. The agricultural policy of the SPDC government is to increase production, but increasing agricultural production means intensive utilisation of land, water and other natural resources, in association with traditional agronomic technologies. This has a direct impact on the condition of soil and water, which, if not properly managed, could lead to environmental degradation. There is no legal or regulatory mechanism to balance the growth of agricultural industries and the increase in environmental impacts.

In addition, increasing use of chemicals and pesticides in the expansion of agricultural production could lead to soil and water pollution. Therefore, environmental problems, further exacerbated because of expected increases in population and food demand, are in the making.

Forestry

In line with the policy of 'protection and conservation of the environment', the SPDC's National Commission on Environmental Affairs (NCEA) initiated forest protection and conservation activities by establishing a Forest Conservation Committee. The NCEA states that one of the main objectives of the Forest Department is to 'manage its forest in such a way that they contribute increased sustained yield and value-added products'. From an estimated forest cover of 500,000 square kilometres, or 70 per cent of Burma's total land area in 1948, the NCEA insists that 50 per cent of the country is still covered with forest (NCEA 1992:12; Smith 1994:12). Most recently, the SPDC claims that 43.3 per cent of total land area of Burma is covered with closed forest.[5]

One of the most visible threats to Burma's environment today, however, is the rapid depletion of many of the country's once great forests. Estimates by independent observers put remaining forest cover in Burma at close to 30 per cent of its total land area. The Rainforest Action Network, for instance, has calculated Burma's annual deforestation rate at 800,000 to one million acres a year, making it one of the five highest in the world (Associated Press 2001). Another independent source puts the rate of deforestation between the regime's claim and that of the Rainforest Action Network's report, stating that forest cover as a percentage of original forest is 40.6 (UNDP et al. 2000). The government-reported data compiled by the World Bank suggest forest cover in Myanmar declined from 60 per cent in 1990 to 50 per cent in 2002 and 49 per cent in 2005. Even though statistical data differ between the government's and independent estimates, the clear message is that Burma's forests are facing degradation at an alarming rate.

The Burmese military government claims that the degradation of forests is due to 'shifting cultivation, local fuel wood shortage, and to a certain extent, the impact of population growth'. According to research sponsored by the United Nations Food and Agriculture Organization (FAO) conducted between 1985 and 1990, the rural population (30.9 million out of 41 million, according to the 1990 census) relied heavily on fuel wood and charcoal for cooking, lighting and heating (United Nations 1992). This trend of fuel wood consumption during the previous decade would have continued at least at the same level between 1990 and 2000, if it did not increase. The military government's claim fails to include the deforestation caused by rapid expansion of the logging trade, which is not reported (Brunner et al. 1998; *The Irrawaddy* 2001; World Rainforest Movement 2002). New commercial contracts were first offered by the regime in late 1988 to neighbouring Thailand (EIA 2002). Many logging companies do not necessarily comply with the logging standards required by the Myanmar government under the system known as the Burma Selection System (BSS),[6] which was created during the colonial period to regulate logging.

The current military regime has enacted a number of laws in order to protect and conserve national forests. The 1992 Forest Law recognises the value of forest beyond commercial uses. It emphasises 'conservation and protection' to meet the needs of the public and the 'perpetual enjoyment of benefits' from the forest (Myanmar 1992). Although the technical competence, skills and commitment of the personnel within the Forest Department are high, these laws, in reality, are no more than window-dressing since top-level officials in the military regime and their cronies are widely believed to be accepting bribes to turn a blind eye to logging companies that often do not comply with the written laws (*Bangkok Post*, 4 January 1998). Therefore, although Burma is not in immediate danger of wiping out its forests in the next few years, current activities, especially intensive logging in eastern and northeastern border areas, are leading to disturbing trends in widespread, and socially destructive, environmental decline. The likely permanent damage to the biodiversity-rich remaining forests is an environmental crisis in the making.

The SPDC's environmental governance

As a policy response and to provide a governance mechanism to address environmental issues, in 1990, the military regime established the NCEA to 'educate the public about environmental awareness'.[7] The NCEA is also charged with formulating a 'comprehensive national environmental strategy' in keeping with a 'modern and developed nation' (NCEA 1992:3). In 1994, the NCEA adopted the National Environmental Policy, which it claims has two major tasks: institutional development, and carrying out the National Environmental Action Plan (Johnson and Durst 1997; FAO 1997:194). The strategies adopted are to: upgrade the NCEA into a statutory body; restructure the NCEA for policy implementation; and achieve financial autonomy for the NCEA. These three objectives demonstrate that the military regime is aware of the need for institutional mechanisms to address environmental issues. An assessment after a decade of institutional development of the NCEA, however, indicates that these objectives have not been fully realised. The first two have been progressing slowly, however, the NCEA is far from achieving financial autonomy because the SPDC has not set up the commission as a statutory body with the formal independent authority to issue policies and implement them.

Acute environmental issues such as forest degradation, water resources management and the sustainability of agriculture come under the authority of the respective departments and ministries that are statutorily separate from the NCEA, which was attached to the Ministry of Foreign Affairs.[8] This arrangement has hindered the development of the NCEA as an implementation body. Therefore, although the language of the NCEA is in tune with the challenges Burma faces in environmental affairs, its current institutional foundations impair its practical effectiveness as a regulatory body.[9]

National environmental action plan

Although the NCEA is charged with drafting a national environmental action plan, it has been moving very slowly. The drafting process of the plan is to focus on: drawing up comprehensive environmental legislation; reviewing and drafting sectoral legislation; conducting environmental impact assessments and establishing environmental standards; collecting environmental data; promoting environmental awareness; alleviating poverty; and setting up sectoral linkages.

The language of the NCEA's draft national environmental action plan demonstrates again that the military regime is aware of the depth and breadth of challenges of national environmental governance. What is happening in reality, however, is different from what is laid down on paper as official policy. In the absence of a national constitution, a national parliament and a legislative body, there is no appropriate institutional mechanism to pass national environmental laws.

Moreover, addressing environmental problems requires input and compliance from different sectors and citizens.

Among all the listed actions in drafting the environmental action plan, promoting environmental awareness has perhaps been the most successful. Achieving some level of environmental awareness among the population could be considered a success, but this success is not meaningful until and unless citizens have material and mental capacities as well as freedom to initiate self-governing community projects and programs that are crucial for successful environmental governance. Governance is in turn closely tied to livelihood issues within the local context.

Since Burma gained independence, the NCEA and its policy framework are the first and only initiatives of their kind designed to address environmental issues in the country. If one examines the reason behind the establishment of the NCEA and the national environmental action plan, however, it is clear that it was driven by external forces, and not by domestic needs or genuine interest in environmental issues. The establishment of the NCEA and the emergence of a national environmental policy in Burma were driven by global awareness and initiatives taken by the United Nations.

The then chairman of the NCEA, U Ohn Gyaw, who was also Foreign Minister, stated

> Burma's commitment and concern for the global and national environment is reflected in the signing of the Framework Convention on Climate Change and the Convention on Biological Diversity at the 1992 United Nations Conference on Environment and Development (UNCED). Environmental protection and conservation occupy a place of special significance on the national agenda of Burma, and Burma's National Commission for Environmental Affairs will continue to strengthen its efforts for preserving and protecting the environment while participating and cooperating in the global effort (U Ohn Gyaw, quoted by Burma Permanent UN Mission Office).

Although Burma has a number of environmental laws and regulations (Table 9.1), it lacks the institutional framework to carry out 'protection and conservation of the environment' so as to achieve sustainable development by implementing these laws, which were not crafted on the basis of sound science or debated democratically to reflect the legitimate livelihood concerns of the population. On top of that, there is no evidence of a political commitment to deal with environmental affairs effectively even under these less than perfect laws. Any careful observer of Burmese affairs would be puzzled by the initial establishment of the NCEA under the Ministry of Foreign Affairs. Why would environmental matters be placed under that ministry—unless it was to showcase a positive image to the outside world. If the regime was serious about tackling environmental matters, it would have established a separate ministry or department and appropriated adequate financial and human resources to tackle environmental matters systematically.

Table 9.1   Current major environmental legislation in Burma

Law and regulation	Year	Purpose
Factory Act	1951	To make effective arrangements in every factory for disposal of waste and effluence, and for matters of health, cleanliness and safety.
Public Health Law	1972	To promote and safeguard public health and to take necessary measures in respect of environmental health.
Territorial Sea and Maritime Zone Law	1977	To define and determine the Maritime Zone, Contiguous Zone, Exclusive Economic Zone and Continental Shelf and the right of the Union of Myanmar to exercise general and exclusive jurisdiction over these zones and the Continental Shelf in respect of preservation and protection of the marine environment, its resources and prevention of marine pollution.
Fishing Rights of Foreign Vessels Law	1989	To conserve fisheries and to enable systematic operation in fisheries with participation of foreign investors.
Marine Fisheries Law	1990	To conserve marine fisheries and to enable systematic operation in marine fisheries.
Forestry Law	1992	To implement forest policy and environmental conservation policy, to promote the sector of public cooperation in implementing these policies, to develop the economy of the State, to prevent destruction of forest and biodiversity, to carry out simultaneously conservation of natural forests

and establishment of forest plantations and to contribute to the fuel requirements of the country.

National Environmental Policy	1994	To establish sound environment policies in the utilisation of water, land, forest, mineral resources and other natural resources in order to conserve the environment and prevent its degradation.
Protection of Wildlife and Wild Plants and Conservation of Natural Areas Law	1994	To protect wildlife, wild plants and conserve natural areas, to contribute towards works of natural scientific research, and to establish zoological gardens and botanical gardens
Myanmar Mines Law	1996	To implement mineral resources policy.
Fertiliser Law	2002	To boost development of the agricultural sector, control fertiliser businesses, and to facilitate conservation of soil and the environment.

Source: United Nations Development Programme, 'The World of Information: Asia and Pacific Review', *The Economic and Business Report*, 1997 Sixteenth Edition. United Nations Development Programme, 1998. *Human Development Report*, United Nations Development Programme, New York.

Another sleight of hand by the regime can be detected in the contents of these major environmental laws, which are broad and do not often have specific standards or regulations to give effect to practical governance of daily environmental issues. Yet the use of separate implementing regulations, or rules, is common practice under the SPDC's rule when it is serious about the implementation of its laws. One should be cognisant of the fact that the creation of the NCEA and announcements about these major laws occurred during the period in which the regime was endeavouring to attract foreign investment and planning to promote tourism in Burma under the slogan 'Visit Myanmar Year 1996'.

Evaluation of the SPDC's environmental governance

Environmental governance in Burma under the SPDC is little more than a façade to present a favourable image on the international stage to encourage tourism and to attract foreign investment. It is one of the usual window-dressing tactics that the military regime has over time mastered for public relations purposes. If the regime was serious and sincere in addressing environmental issues, it would first need to address the fundamental problems of local people's livelihoods and freedom of entrepreneurial activity. Pressing environmental issues include public health, sanitation, clean drinking water, soil erosion, agricultural technological development, assessing the impact of importing foreign seeds (Phyu 2006) and proper designs for irrigation projects. To address these issues effectively at the national level, the regime has to appropriate funds and establish an independent body to coordinate (not consolidate) activities among different ministries, ensure citizens' active participation, ensure information about development projects is shared and conduct environmental and social impact assessments by commissioning independent scientists and experts from local communities. Since none of these are practised in the SPDC's Myanmar, trying to measure the state of environmental governance by using the normal standards in project implementation and impact assessments is like 'shooting a sparrow with a rocket launcher', as the Burmese saying goes.

Most important of all, the freedom of local farmers and communities to make sound judgments and decisions about their livelihood issues, which are directly associated with utilisation of natural resources, has to be honored. This is unlikely to happen, however, in the current mind-set of the SPDC, which is intolerant of new ideas at the community level and short of vision for meaningful development of the country or for improving the well-being of the people. The SPDC is entirely preoccupied with restricting political action by the people; it even restricts the basic freedom of farmers to cultivate whatever crops they desire on their land. Therefore, if the SPDC is sincere about addressing environmental governance in Burma, it should at least guarantee freedom for farmers and the local population to pursue their livelihoods and entrepreneurial activities as they choose.

Continuing political crisis and environmental issues

Any study of the issues relating to Burma must pay attention to what is happening on the political stage and its history. Political instability and the fragile state of governance are the crucial problems that have contributed to Burma becoming one of the less developed countries among UN members. Indeed, Burma is perhaps one of the strongest examples of how lack of political development has hindered economic and social progress. No period in post-independent Burma's history has witnessed such impediments to the country's potential social and economic progress as the present time, with its continuing political crisis.

Because of the absence of good governance and appropriate institutional mechanisms to provide checks and balances in the exploitation of natural resources, Burma is beginning to encounter experiences similar to countries that pursue material development ahead of sustainable and equitable social development. It will soon follow the trends of other societies in mainland Southeast Asia, such as Thailand, where original natural forests in lowland areas have been turned into rice paddies, fruit orchards, infrastructure developments and golf courses, and where forest animals are traded in tourist-crowded local markets.

Table 9.2 International environmental conventions ratified or signed by Burma

International environmental conventions	Year in force
Plant Protection Agreement for the Southeast Asia and Pacific Region	1959
Treaty Banning Nuclear Weapons Tests in the Atmosphere, in Outer Space and Under Water	1963
Outer Space Treaty: Treaty on Principles Governing the Activities of States in the Exploitation and Use of Outer Space including the Moon and other Celestial Bodies	1970
Convention on the Prohibition of the Development, Production and Stockpiling of Bacteriological and Toxic Weapons, and their Destruction	1972 (signed)
MARPOL: International Convention for the Prevention of Pollution from Ships 1973	1988
MARPOL Protocol: Protocol of 1978 Relating to the International Convention for the Prevention of Pollution from Ships 1978	1988
Agreement on the Networks of Aquaculture Centres in Asia and the Pacific Region 1988	1990
Treaty on the Non-Proliferation of Nuclear Weapons	1992
ICAO: ANNEX 16 to the Convention on International Civil Aviation Environmental Protection Vol.I, Aircraft Noise	1992
ICAO: ANNEX 16 to the Convention on International	

Civil Aviation Environmental Protection Vol.II, Aircraft Noise	1992
Vienna Convention for the Protection of the Ozone Layer	1994
Montreal Protocol on Substances that Deplete the Ozone Layer (Montreal Protocol)	1994
London Amendment to the Montreal Protocol	1994
Convention on the Prohibition of the Development, Production and Stockpiling and Use of Chemical Weapons, and their Destruction	1994
Convention Concerning the Protection of the World's Cultural and Natural Heritage	1994
Framework Convention on Climate Change (FCCC)	1995
Convention on Biological Diversity (CBD)	1995
United Nations Convention on the Law of the Sea	1996
International Tropical Timber Agreement (ITTA)	1997
United Nations Convention to Combat Desertification	1997
Convention on International Trade in Endangered Species of Wild Fauna and Flora (CITES)	1997

Source: Author's compilation

Local rights in national development

One of the fundamental challenges Burma must address, if sustainable development is to be achieved, is the issue of local people's rights and political freedom to manage the natural resources on which their livelihoods are based. When the globalisation of economic activities intensifies pressure on a country such as Burma—where there is no solid foundation of rule of law—the most vulnerable victims are the local people and the natural environment. Within this context, communities and landscapes under the greatest environmental threat today are generally those inhabited by the most vulnerable members of society, including the poor, ethnic minority groups, women, children, refugees and other internally displaced people. Indeed, the lives and living conditions of rural populations along Burma's borders and within the country are examples of this phenomenon.

The continuing political instability in Burma and political corruption in the regional context of the Mekong region and Southeast Asia have provided havens for human rights abuses and often-neglected social problems, which in turn continue to haunt environmental arenas in these countries. Burma's political instability does not contribute to the long-term development of good governance in the Mekong region.

Burma in the context of international environmental governance

What positive steps can be taken from the SPDC's show of environmental governance? The statement by U Ohn Gyaw and careful assessment of the SPDC's policy for the NCEA signal that the military regime often responds to external factors and forces in its own way. After understanding what internal and external factors SPDC take into account in responding to both domestic and international pressures, strategists then can use SPDC's responses as baseline pressure points to build up further factors or standards to which the SPDC will have to make further responses. For instance, when a country signs international treaties and laws (see Table 9.2), signatory states are supposed to be accountable to report on the implementation of treaty clauses and legal frameworks they have set up. Accountability mechanisms can be achieved by applying legal and policy-program means. One of the most successful examples of how the international community deals with the SPDC's responses to external forces is the work of the International Labour Organization (ILO) on forced-labour issues. Perhaps this model can be used to address environmental issues while developing environmental governance.

For example, Burma has signed a number of the international environmental conventions: it has signed, and acceded to, or ratified the Convention on Biological Diversity (1994), the Convention on International Trade of Endangered Species (1979), the International Tropical Timber Agreement (1996) and the Framework Convention on Climate Change (1994) (Table 9.2). It has also participated in the UN Conference on Environment and Development, and received funds through the Global Environment Facility. Burma's path is leading towards increased international engagement in environmental arenas. This engagement can open up channels of communication to discuss environmental issues with the military government. The regime has shown, through its limited environmental initiatives, a 'greening' in some of its policies. Although it can be argued that, ultimately, the regime's policies are merely lip service, the regime has at least demonstrated some level of awareness of environmental issues in Burma.

Moreover, in 1997, Burma became a member of the Association of Southeast Asian Nations (ASEAN), which is leaning increasingly towards regional cooperation in dealing with environmental problems. For example, in September 1997, ASEAN members signed the Jakarta Declaration on Environment and Development and pledged to use resources efficiently and sustainably. As a result, ASEAN set up the ASEAN Regional Centre for Biodiversity Conservation with the aim of supporting and empowering communities to achieve their eco-efficiency objectives. The Mekong River Commission (MRC), with its pre-eminent role in the Mekong region and expanding work program, is another transnational institutional mechanism that can work with the current regime in Burma to establish much needed baseline data and information about the true state of the environment in Burma. Using the SPDC's desire to gain international recognition, the MRC should approach Burma to become a fully fledged member. ASEAN, the MRC and the international environmental treaties that the military regime has signed are all potential institutional mechanisms that can be applied to engage with the regime. Such engagement can at least be aimed at information sharing and dissemination about the current state of environmental affairs in Burma, perhaps leading to training relevant officials for environmental assessment.

Limited opportunities for environmental governance

If proper policies can be crafted and implemented in Burma—ideally within a democratic system of governance—the rich cultural and ethnic diversity along with the natural environmental endowment can be great resources for development of the country. The history of Burma's parliamentary democracy period provides ample empirical evidence that if Burmese people have a certain level of political freedom and stability, they can put the country back on the path to development. Because some level of democratic governance existed from 1948 to 1960, Burma in the 1950s was one of the most promising and respected countries in the region, even though this was the early days of its independence. What was crucial then was a level of political freedom for governmental agencies and citizens to implement innovative ideas in their daily livelihoods without unjust restrictions from an authoritarian government. Can environmental issues be applied to create political space for self-governance and more freedom for local populations, and in turn contribute to political transition as a whole? This question should be a guiding one that feeds into the continuing approaches taken by local and international environmental NGOs on environmental issues and governance in the current political crisis in Burma. This is fundamental for environmental governance, which depends ultimately on the inputs of the local population.

Cases of limited environmental success

Although this chapter emphasises a bleak view of the current political context and governance structures, it also acknowledges that some limited opportunities exist in Burma for environmental governance. Some of the existing mechanisms that show a level of success and are appropriately addressing the importance of environmental issues in Burma are discussed below.

Smithsonian Institution

The Smithsonian Institution has been working with the Forest Department in Chatthin Wildlife Sanctuary (CWS) since 1992 after selecting the sanctuary as a site for a multi-year ecological study (Aung et al. 2004). The goal of this project is to build local capacity in order to secure the future of the ecosystem. This goal is accomplished through training the sanctuary staff, conducting ecological research and fostering community-based conservation. From 1992 to 2000, training sessions on birds, mammals, herpetology and entomology inventories, community relations and environmental education were offered. The strength and prospects of this project, however, rely on the good grace of the Forest Department. A simple change of leadership in wildlife sanctuary superintendent could risk the future of the project and it could be unnecessarily delayed. There is a lack of institutional structures to support such a project for the long term, even though it could survive in the short term through management of 'personal diplomacy'. The Myanmar government sees these projects as window-dressing opportunities to improve its international reputation and gain much desired legitimacy—from 'less political' issues, such as the environment.

It would be wise to assume, however, that the military government does not want this type of project to expand too deeply into other wildlife sanctuaries or environmental concerns. The perceived reason is that these externally run projects risk giving the impression that the government is incapable of organising and managing such projects without outside intervention. The regime is constantly claiming that it is protecting Burma's sovereignty and independence from outside influences. Therefore, even though a project like this can be initiated and survive to a degree under the leadership of the relevant ministry, such projects face an uncertain future until a proper political settlement is achieved. Nevertheless, if these projects continue to operate within the workable framework at present, they will contribute enormously to environmental governance once the country becomes an open and democratic society.

Wildlife Conservation Society

Rao et al. (2002) reported on the Wildlife Conservation Society's (WCS) remarkable efforts on assessing 22 of the 31 official protected areas in Burma. Such a scientific assessment of the state of the environment in Burma is needed but can rarely be undertaken. The WCS, however, with the assistance of Forest Department personnel, was able to undertake this important assessment of protected areas and produce empirically grounded strategies to help strengthen current conservation efforts. Meanwhile, the WCS also worked closely with the Forest Department in the creation of the new 3,812 square kilometre Khakaborazi National Park.

The WCS-led study found that grazing, hunting, fuel wood collection and permanent settlements occurred in more than 50 per cent of the protected areas surveyed, with biodiversity loss most severe in older protected areas and less severe in the newly created national parks, such as Khakaborazi (Rao et al. 2002:364). The study issued eight recommendations including building the technical capacity of protected areas staff, involving local communities in protected area management, implementing a comprehensive land-use plan, controlling hunting and amending wildlife laws to fulfill international treaty obligations. This type of assessment could be adapted to other areas of environmental concern, such as the management of forests, rivers, lakes and wetlands.

United Nations Development Programme Watershed Project Initiatives

The United Nations Development Programme (UNDP) has been engaged in at least three continuing projects as a part of its Human Development Initiative. These projects are located in the Dry Zone, the Ayeyarwaddy Delta and in Southern Shan State. They aim to promote environmentally sustainable practices, food security and micro-income opportunities. One of the success stories of the UNDP project is in Southern Shan State, where deforestation, shifting cultivation, poverty, over-grazing and forest fires are constant problems. The Southern Shan State project relies heavily on community forestry initiatives to enable communities to regain control of their forests, feel a sense of ownership and promote true responsibility in taking care of the forest. Since the project's inception in 1994, 764 acres have been accepted as community forest and another 1,335 acres have been reported as pending acceptance. Although the number of acres under community forests is small, the UNDP project has identified a total of 306,516 acres as potential community forests (Sterk 1999).

The UNDP has wider political acceptance and legitimacy in the eyes of the government and people in Burma. At the same time, the status and image of the United Nations and its agencies are relatively sensitive and more open to criticism by the international community, NGOs and the Myanmar opposition, some of whom advocate isolating the military regime. Therefore, the dilemma of aid from UN agencies is more intimately tied to the political context of Burma than those of independent international agencies, such as the Smithsonian Institution and the WCS, which could choose to aid Burma regardless of criticism.

One has to be mindful of the fact that although these opportunities exist and they can be used to initiate environmental governance in Burma, it is difficult to predict the potential outcomes and whether these projects can sustain the momentum for the long term. If these projects—aimed indirectly at contributing to political transition by creating political space for the local population and disseminating information about livelihood conditions and the environment—continue, they will make positive contributions to environmental governance in Burma. But as long as the regime refuses to commit to a transition to democracy supported by the Burmese people and the international community, these projects face an uncertain future.

Challenges and strategies for environmental governance

The initial challenges for environmental governance in the SPDC's Myanmar are rooted in three dimensions: institutional development; budget or resource capacity; and knowledge or environmental education (capacity building). Institutional development for environmental governance is hindered by many factors, including the lack of political will and the continuing political crisis. Resource capacity and budget funding for environmental governance are at the bottom of the list of the military regime's priorities. There is no official report from the regime as to how much of the annual budget is appropriated for environmental governance, such as for building infrastructure or monitoring and enforcing environmental laws and training staff. According to the WCS's survey report, '[N]one of the protected areas surveyed had the necessary infrastructure for effective reserve management or sufficient on-site personnel to perform park management activities adequately' (Rao et al. 2002:364). Environmental education of the general populace and among staff at the relevant ministries is another challenge for environmental governance. For instance, Rao et al. (2002:363) reported that only 35 per cent of the national parks surveyed had approximately half of their staff trained in basic field techniques. If one of the most active areas of environmental governance—national park or protected area management—has such a small number of staff trained systematically, one can imagine the low level of environmental management education in other areas. Therefore, the first challenge for environmental governance in Burma is to understand the depth and breadth of the challenges that lie in these three dimensions.

Strategies

In a broader sense of promoting good governance at the national level, one possibility is for the international community to put pressure on the regime to adhere to the rule of law and to provide a future scenario for Burma's environmental governance. As Burma has signed a number of important international environmental conventions, these can be used to monitor its environmental affairs. At the same time, these conventions provide mechanisms for international environmental organisations to engage with the military regime and test its willingness to abide by international standards. International environmental NGOs and UN agencies can also engage in *ad hoc* training in environmental law related to the conventions to which Burma is a party. This approach can be initiated with projects to train relevant government officials to understand these international laws and treaties. Such an approach might be carried out by a credible international NGO that has the capacity and a genuine interest in Burma's environmental future.

Another channel for launching this type of approach might be some mechanism under ASEAN or the MRC, both of which could provide training courses on international environmental laws for relevant government officials. For instance, training government officers in the Forest Department about international standards for protected area classification, field techniques, survey methods and environmental education in general could lead to the creation of more protected areas in Burma. After testing the regime's seriousness to adhere to international environmental conventions, specific strategies that are in line with the 1994 National Environmental Policy could be applied to support the institutional development of the NCEA.

First, institutional development of the NCEA to become a separate statutory body for environmental policy and governance might be accelerated by partnership with an international organisation such as the UNDP, the UN Environment Program (UNEP), the Global Environmental Facility or an international donor agency acceptable to the regime. A potential international funding organisation or agency might approach the regime by offering a loan or aid with an agreement to establish a ministry or department of the environment. The agreement should clearly outline workable mechanisms and functional autonomy for the ministry or department to implement domestic and international environmental laws and policies. Ideally, a UN environment-related agency should take up such an initiative, because the military government's trust in such an agency is likely to be much higher than in a country-based international environmental NGO. It is to be expected, however, that Burmese opposition forces would criticise such a project. If such a project were able to adopt a political approach, the agreement should clearly state that the monitoring board for the institutional development of the NCEA must be composed of representatives from the regime, funding agencies and the democratic opposition led by Aung San Suu Kyi. Such an approach would avoid unwanted criticism from the opposition. Perhaps, such a partnership project could contribute to UN-initiated dialogue between the regime and the opposition.

Second, applying Conservation International's Guyana model of 'conservation concession' would generate the potential to increase more conservation areas in Burma. It is reported that only 2.26 per cent of the total area of the country is designated as protected areas (Rao et al. 2002:361). Conservation International obtained the first conservation concession in 2000 from Guyana, a small former British colony on the north coast of South America. A concession is a lease on a parcel of land granted by a government for a specific purpose. In the Guyana case, Conservation International leased a 200,000-acre tract in the remote southeastern corner of the country for an application fee of US$20,000 and 15 cents per acre annually. Conservation International then put up additional funds for management of the tract as a nature reserve. The initial period was for three years during which both parties would negotiate the rate for a subsequent 25 years (Wilson 2002). This model can be applied to increase conservation areas in Burma, however, this requires strategic selection of biodiversity-relevant sites of global importance within Burma and an assurance of enforcement mechanisms from the military regime.

Finally, to prevent the illegal trade of wildlife and timber in border areas, there is a need for trans-border cooperation between Burma and neighbouring countries—because wildlife protection and deforestation in Burma are driven largely by international and cross-border demands. Trans-border measures should be promoted, especially among Burma, China and Thailand. The emergence of such measures will depend on the political commitment of neighbouring countries. First, it will require an appropriate platform to address the issue. One place to start would be through the Greater Mekong Subregion (GMS) development scheme supported by the Asian Development Bank (ADB). If ADB funding for the development of roads and energy networks in the GMS imposed trans-border environmental conditions, it could succeed in at least the establishment of an official trans-border environmental coordination committee. Another model of trans-border cooperation is the Haze Technical Task Force set up by ASEAN to solve subregional issues. Although the Haze task force was not successful because it lacked an operational agenda, it at least provides a model to start dialogue on subregional trans-border issues, such as wildlife and illegal timber trading.

Conclusion

With its continuing political instability, war and repression, Burma stands to lose much of its remaining natural resources at an alarming rate. The military regime's protection and conservation of natural resources and the environment as a 'national endeavour' has been couched in progressive language. The drafting and implementation of its National Environmental Policy is, however, yet to produce appropriate institutional mechanisms. Any strategic environmental engagement with the military regime will have to bear in mind that a fruitful result for sustainable environmental governance in Burma, and consequently in the ASEAN and Mekong regions, will depend on the existence of good governance practice in a broader sense. Transparency, accountability, rule of law, an independent judiciary system and mechanisms to include local participation in environmental decision making are essential for good governance practices. Burma lacks most of these elements, although there are some limited possibilities for local participation, as can be seen from the success of the UNDP's projects. Therefore, until and unless national reconciliation is reached and political differences are resolved among all concerned parties, Burma's environmental future will be held hostage by political instability. It is desirable that the short-term successes of the projects discussed in this chapter lead to the rescuing of the hostage.

It is crucial that the leaders of the SPDC regime realise that the existence of human civilisation depends inevitably on the harmonious relationship between society and the environment. The common finding of scientists who study the reasons behind the survival and collapse of earlier civilisations is that those civilisations collapsed due to a lack of vision and a lack of institutional arrangements to achieve a balanced relationship between society and the environment (Hodell et al. 1995; Weiss and Bradley 2001; Haug et al. 2003). The great lesson that the SPDC generals can learn from the collapse of states in the past is that the meaningful development of a society and the continuing existence of a civilisation depend on human ideas, capacities and political freedom within that society. Burmese society is endowed with ideas and capacities; what is lacking is political freedom for citizens to exercise their ideas and capacities. If current political deadlocks continue to deny citizens the political freedom to chart their own livelihoods and self-governance into the future, Burma's civilisation and its continued existence in the modern context will be at risk. This assessment of environmental governance under the SPDC would have to conclude that the primary responsibility for charting better environmental governance in Burma lies in the hands of the SPDC generals.

Notes

1   The SPDC uses this term in its propaganda campaign against the opposition.
2   Then Lieutenant-General Khin Nyunt, Secretary One of the SPDC, stated on 15 May 1991 in an interview that martial law meant 'no law at all'. See Amnesty International 1992.

3 Élitist Burmese are those who have access to and cozy relationships with military leaders although they cannot be considered as SPDC regime elements.
4 See Mirante 2002. For logging, see Brunner et al. 1998. See also *The Irrawaddy* 2001 and World Rainforest Movement 2002.
5 see http://www.energy.gov.mm/MOF_1.htm
6 The Burma Selection System requires recording the age of trees. It involves a 30-year felling cycle based on minimum size selection criteria. See http://www/Burma.com/gov/perspec/
7 Major General Khin Nyunt, then Secretary One of the State Law and Order Restoration Council (SLORC) and head of Military Intelligence, 14 May 1991, quoted in *The Working People's Daily*, 16 May 1991.
8 Fifteen years after the creation of the NCEA, the SPDC in 2006 removed the NCEA from the Ministry of Foreign Affairs to the Forestry Ministry.
9 It is unlikely that the NCEA can become a statutory body under the SPDC without a proper constitution and legislature.

References

ABSDF Research and Documentation Center Office, (unpublished). *Yawn Sone Tatmadaw (Colourful Army)*, ABSDF Research and Documentation Center Office, Thailand.
Amnesty International, 1992. *Burma: no law at all: human rights violations under military rule*, October, Amnesty International, London.
Associated Press, 2001. 'Working elephants make last stand in Myanmar's teak forests', Associated Press, 3 March 2001.
Aung, M., Swe, K.K., Oo, T., Moe, K.K., Leimgruber, P., Allendori, T., Duncan, C. and Wemmer, C., 2004. 'The environmental history of Chatthin Wildlife Santuary, a protected area in Myanmar', *Journal of Environmental Management*, (72):205–16.
Bangkok Post, 1998. 'Thai–Burmese illegal logging involves influential people', *Bangkok Post*, 4 January.
Brunner, J., Talbott, K. and Elkin, C., 1998. *Logging Burma's Frontier Forests: resources and the regime*, World Resources Institute, Washington, DC.
Bryant, R.L., 1993. 'Forest problems in colonial Burma: historical variations on contemporary themes', *Global Ecology and Biogeography Letters*, 3(4/6):122–37.

Environmental Investigation Agency, 2002. *Logging in SE Asia and international consumption of illegally sourced timber*, Environmental Investigation Agency and Telapak Indonesia. Available from http://www.eia-international.org/Campaigns/Forests/Reports/timber/timer04.html

Food and Agriculture Organization, 1995a. *Implementing Sustainable Forest Management in Asia and Pacific*, Food and Agriculture Organization:183–95.

——, 1995b. *Review of Wood Energy Data in REWEDP Member Countries*, Food and Agriculture Organization:52–5.

——, 1997. *Country Report—Myanmar*, Food and Agriculture Organization, Rome.

Gyaw, O., 1997. Statement of Chairman of NCEA to Nineteenth Special Session of the UN General Assembly, 24 June 1997.

Haug, G.H., Günther, D., Peterson, L.C., Sigman, D.M., Hughen, K.A. and Aeschlimann, B., 2003. 'Climate and the collapse of Maya civilization', *Science*, 299:1,731–5.

Hodell, D.A., Curtis, J.H. and Brunner, M., 1995. 'Possible role of climate in the collapse of the Classic Maya civilization', *Nature*, 375:391–4.

Johnson, A. and Durst, P.B., 1997. 'Implementing sustainable forst management in Asia and the Pacific. Proceedings', FAO no. 1997/7.

Lintner, B., 2002. 'China and South Asia's East', *Himal South Asian: Burma Special*, October.

Major General Khin Nyunt, 1991. Quoted in *The Working People's Daily*, 16 May.

Mirante, E.T., 2002. 'Gunboat petroleum: Burma's Unocal/Total pipeline', *Environmental News Network*, 26 April.

Myanmar, 1992. *Myanmar Forest Law*, No.8/92:Chapter II Basic Principles.

Myanmar, 2007. Available from http://www.energy.gov.mm/MOF_1.htm.

National Commission for Environmental Affairs of Burma, 1992. *Report to UNCED*, June, National Commission for Environmental Affairs of Burma:5–12.

Phyu, K.H., 2006. 'Inle farmers call for the use of local seeds', *Myanmar Times*, 18 May.

Ponting, C., 1991. *A Green History of the World: the environment and collapse of great civilizations*, Penguin Books, London.

Rao, M., Rabinowitz, A. and Khaing, S.T., 2002. 'A status review of the protected area system in Myanmar with recommendations for conservation planning', *Conservation Biology*, 16(2):360–68.

Smith, M., 1994. 'Paradise lost?: suppression of environmental rights and freedom of expression in Burma', *Article 19*, September:9–12. Available from http://www.article19.org/pdfs/publications/myanmar-environment.pdf.

Sterk, A., 1999. *Consultancy Mission Report on Community Forestry—Union of Myanmar*, United Nations Development Programme and Food and Agriculture Organization, Bangkok.

The Irrawaddy, 2001. *The Irrawaddy*, 9(8), October–November. Available from http://www.irriwaddy.org/database/2001/vol9.8/cover.html.

United Nations, 1992. *Sectoral Energy Demand in Burma*, Regional Energy Development Program, United Nations, Bangkok.

United Nations Development Programme, 1998. *Human Development Report*, United Nations Development Programme, New York.

United Nations Development Programme, United Nations Environment Program, World Bank and World Resources Institute, 2000. *World Resources 2000–2001: people and ecosystems: the fraying web of life*, United Nations Development Programme, United Nations Environment Program, World Bank and World Resources Institute, Washington, DC. Available from http://pubs.wri.org/biodiv/pubs_description.cfm?pid=3027 (accessed 4 June 2007).

Weiss, H. and Bradley, R.S. 2001. 'What drives societal collapse?', *Science*, 291(5504):609–10.

Wilson, E. O., 2002. *The Future of Life*, Alfred A. Knopf, New York.

'The World of Information: Asia and Pacific Review', *The Economic and Business Report*, Sixteenth Edition, 1997.

World Rainforest Movement, 2002. *Bulletin*, No. 54, January. Available from http://www.wrm.org.uy/bulletin/54/Burma.html

10 Environmental governance of mining in Burma

Matthew Smith

Mining is the extraction of non-renewable resources; as such, it is an inherently unsustainable practice. Even when carefully managed and monitored, mining always has social and environmental costs. This is especially true in developing countries, where environmental governance tends to be weaker than in industrial countries. Burma is an authoritarian state that has been ruled by successive military governments since 1962. The human rights violations and environmental degradation around the mining industry in Burma are similar to those happening in other extractive industries in the country, and they are indicative of the state of environmental governance: unfair and inefficient.

This chapter is an analysis of environmental governance of mining in Burma. I argue that it is a top-down system, devoid of environmental protection and dominated by the elemental purpose of securing revenue. While far short of an exhaustive analysis of the environmental governance of mining in Burma, this chapter provides a discussion of the varieties of mining in Burma, challenges to environmental governance analysis in Burma and the national economic policy and mining laws, illustrated through a case study of the Monywa Copper Project in Sagaing Division. This project is Burma's largest mine and a joint venture between the Canadian company Ivanhoe Mines Limited and the Myanmar Ministry of Mines under the State Peace and Development Council (SPDC).

This case study considers the structure of the Monywa Copper Project, the methods and processes used at the Monywa mine site, recent developments concerning Ivanhoe Mines' investment in Burma and the highly destructive artisanal mining—called *dohtar* in Burmese—that is widespread around the Monywa mine site. Highlighting the weaknesses and problems with environmental governance and mining in Burma begs questions of how the situation might improve. In the final section, I emphasise areas in need of attention, as well providing a forecast for local participation in fair and effective environmental governance in Burma.

Challenges to environmental governance analysis

Mining in Burma is widespread and conducted in various ways. In 2004, the Ministry of Mines reported 43 large-scale mining permits, 165 small-scale permits and 1,320 subsistence permits (ABARE and Mekong Economics 2005). As of 2005, many of these permits were inactive, which is not to say that the mining industry in Burma is inactive or that the Ministry of Mines' figures are accurate.

For the sake of order, mining in Burma can generally be categorised as large-scale, artisanal and small-scale mining. The latter two types of mining are often grouped together—as 'ASM'—by researchers due to the commonalities between them.

Large-scale mines such as the Monywa Copper Project are enormous operations, and are often financed by foreign investors. They have high recovery and production levels and use technologically advanced methods and equipment, including heavy machinery and complex chemical processes.

Small-scale mines vary in size, are more labour intensive than large-scale mines and may or may not use mechanised equipment. In Burma, small-scale mines are run by the military and non-military alike.

Artisanal mining is characterised by rudimentary, traditional methods, is labour intensive and occurs informally, always as a means of subsistence. ASM is driven by poverty; it requires little or no capital inputs and its accessibility makes it pervasive.

Large and small-scale mines operate in official capacities in Burma with mining permits from the State or a partner of the State, while artisanal miners are viewed as illegal, often working on land without legal title. All types of mining are characterised by their significant impact on the environment (Images Asia and PKDS 2004).

In Burma, these three types of mining are also characterised by difficulties in collecting quantitative and/or qualitative data sets pertaining to them, in turn making questions of environmental governance more challenging, and more pressing. Large-scale mining companies have a global reputation for secrecy around their operations and exploration and, when partnered with the SPDC in a closed society, the documentation of mining and its effects becomes particularly challenging. This is an initial area of concern for environmental governance around large-scale mining in Burma: lack of transparency.

As for ASM in Burma, the initial challenge in terms of environmental governance analysis is less about transparency—because artisanal mining occurs in an unofficial capacity—and more about surmounting the obstacles to data collection. Much of the economic activity around ASM is never reported officially, making any wide documentation of it and its effects particularly challenging, if not impossible, under the current military regime (EarthRights International 2003).

It is estimated that more than 13 million people world-wide depend directly on ASM for survival, and a further 80–100 million people's livelihoods are affected by ASM (Ayers et al. 2002; Hentschel et al. 2003). In Burma, national figures of the number of ASM miners and the environmental and social impacts of their activities are difficult to estimate beyond generalising it as widespread and pervasive. But consider that mining in Burma occurs nation-wide, and the Ministry of Mines officially supports the mining of copper, lead, silver, zinc, refined tin, tin concentrates and tin-tungsten, gold, iron, steel, coal, and the production of industrial minerals, as well as gems and stones, pearls and salt (Myanmar Ministry of Mines n.d.). Many of these formal mining enterprises have an artisanal shadow: an informal ASM sector working alongside them.

The most formidable challenge to environmental governance analysis in Burma is, essentially, access to information. A lack of access to information pervades Burma's environmental governance system and is perhaps indicative of a system more concerned with generating and securing revenue than with collecting and reporting useful information about the state of Burma's environment. Indeed, to speak of challenges to environmental governance analysis is very different than to speak of challenges to environmental governance, though of course the two are closely interrelated. The following sections explore some political dimensions of the governance problem, starting with Burma's national policy and mining laws.

National policy and mining laws

A comprehensive analysis of Burma's environmental governance system, for better or worse, is a task well beyond the general scope and specific limits of this chapter. That said, Burma's National Environmental Policy (NEP) and mining laws are two essential indicators of the state of environmental governance and mining in Burma, and are both worthy of a narrow and specific focus.

The National Commission for Environmental Affairs (NCEA) was created in Burma in 1990, followed closely by the drafting of the NEP. The NEP invokes the universally agreed principle of sustainable development, concluding as follows: 'It is the responsibility of the State and every citizen to preserve its natural resources in the interests of present and future generations. Environmental protection should always be the primary objective in seeking development' (NEP cited in Tan 1998). While a noble goal indeed, most of the NEP reads like a mission statement, or a disjointed collection of environmental platitudes. For example, it reads: 'The wealth of the nation is its people, its cultural heritage, its environment and its natural resources', continuing to say that 'Myanmar's environmental policy is aimed at achieving harmony and balance'. Considering the well-documented environmental irresponsibility, degradation and destruction happening in connection with Burma's extractive industries, and the near complete absence of institutional capacity, the policy rings hollow and sounds generalised.

The policy, however, is not without foundations. It does demonstrate an unsurprisingly selective expression of international environmental law, clearly noting the country's right of permanent sovereignty over its own resources, which is consistent with the SPDC's general insistence on sovereignty and its contempt for international pressure, which is deemed a threat to that sovereignty. The policy states that: 'Every nation has the sovereign right to utilize its natural resources in accordance with its environmental policies', which is taken almost verbatim from Principle 21 of the Stockholm Declaration, which describes nations as having 'the sovereign right to exploit their own resources pursuant to their own environmental policies' (UNDP 1972; UNEP 1992). Numerous other expressions of international environmental law beyond state sovereignty exist that would succeed in bringing the NEP in line with international environmental legal standards. However, the real concern with the NEP is less about its content and more about Burma's institutional capacity, which is unable to give effect to the policy and its stated environmental maxims. Until that capacity is in sight, with democratic preconditions in place, the NEP will continue to merely espouse, as opposed to enable, sustainable development.

While the NEP espouses sustainability and environmental protection as the primary objectives of development, on its web site, the Myanmar Ministry of Mines ostensibly notes its own elemental purpose: 'It is the policy in the mineral sector to boost up present production, to fulfil the growing domestic demand and to increase foreign exchange earnings.' The aim of the Ministry of Mines to secure revenue is often wholly at odds with the protection and conservation of the natural environment.

Mining natural resources can be a highly productive economic activity. It is undertaken to generate profit, and as such the stakeholders directly involved in the enterprise are, quite expectedly, economically interested. These economic interests will want to be protected from measures that might render the enterprise less economic, or worse, uneconomic. This puts sound mining laws and regulations on a collision course with the core intent of mining operations, which is maximising profit.

In light of Burma's authoritarian state, it is reasonable to expect that official top-down measures will be taken to ensure that the large-scale mining of non-renewable natural resources remains economic as opposed to uneconomic or less economic. Measures such as keeping regulations voluntary and specific requirements and duties of private companies minimal are the expected norm in Burma for the foreseeable future.

I turn now to consider the weaknesses of Burma's mining laws, which demonstrate an absence of fair and efficient environmental governance as well as a perverse environmental governance strategy: by their weakness and lack of enforcement, the mining laws facilitate economic activity, maximising the wealth of some stakeholders to the detriment of the least advantaged stakeholders: local people and the natural environment.

Burma's mining laws

The Ministry of National Planning and Economic Development dictates national economic policy in Burma. Large-scale mining and the national economy of Burma are related by legislation in that the SPDC controls the flow of all foreign direct investment in Burma and the SPDC is necessarily a partner in all mining investments (either production sharing or profit sharing).

In 1988, three months after the crack-down on the nation-wide pro-democracy uprising, the SPDC's precursor, the State Law and Order Restoration Council (SLORC), passed Law 10/88, which opened Burma's economy to foreign investment in order to promote development of the national economy. In practice, this policy enables the SPDC to control the flow of foreign direct investment coming into Burma, and shareholding capacity has been reserved for the military and their families (The Burma Campaign [UK] 2004). Since 1988, total foreign investment in Burma is estimated at US$7.646 billion, with mining investments accounting for US$534.19 million, most of which comes from the Monywa Copper Project. This makes the mining industry the fifth most lucrative for the SPDC, behind oil and gas, manufacturing, livestock and fisheries, and real estate (*The Irrawaddy* 2005a).

The Myanmar Ministry of Mines directs the formal mining sector in Burma, and all mining contracts must be approved by the State. This gives an air of cohesion and legislative structure to the national layer of governance. The SPDC maintains that 'all naturally occurring minerals found either on or under the soil of any land on the continental shelf are deemed to be owned by the state'. The Ministry of Mines comprises various branches tasked with granting mining concessions and investigating potential mineral deposits, but as a matter of policy the SPDC has affirmed to refrain from making new mining investments on its own, instead looking to encourage foreign and local investment.[1]

In turn, foreign and local investment in the mineral sector is enabled only by mining concessions awarded by the Ministry of Mines or by close partners who have been given concession authority.[2] Since 1988, when the government acted decisively to 'develop [the] national economy', almost nothing is known about the many official and unofficial mining concessions granted to local and Chinese companies, many of which are owned by or closely tied to armed groups throughout the country (MacLean 2004).

The SPDC has attempted to make foreign investment in Burma attractive, ensuring potential investors that it would not nationalise the industry or the investment for the life of the contract (SLORC 1988). Chapter VI, Article 22 of the Myanmar Foreign Investment Law states that '[the] Government guarantees that an economic enterprise formed under a permit shall not be nationalised during the term of the contract or during an extended term, if so extended' (SLORC 1988). The last article of the same law, however—Chapter XV, Article 32—adds a question to that assurance, stating that for 'the purpose of carrying out the provisions of this Law the Government may prescribe such procedures as may be necessary, and the Commission may issue such orders and directives as may be necessary' (SLORC 1988).

Burma's mining laws are vague and incoherent. They consist largely of general statements lacking the clarity and cogency normally expected of well-written laws. They have been described as 'among the least developed, or sound, of any in the world' (Moody 1999:12, footnote 73 in Chapter IV). Some laws conflict with one another; others are simply redundant (Moody 1999; SLORC 1994). In regard to land rights, Chapter V, Article 15 of the 1994 Mines Law gives legal go-ahead for the government's standard practice of land confiscation, citing 'the interest of the State': 'If, in the interest of the State, it is necessary to acquire the land where mineral production could be undertaken on [a] commercial scale, the Ministry shall co-ordinate with the relevant Ministry for the acquisition of such land in accordance with the existing Law.'

This law in effect means that if you happen to be living in the wrong place—that is, above a mineral deposit—your land is subject to seizure. There is no provision for compensation or even a vague resettlement plan.

Environmental impact assessments (EIAs) are in one way or another becoming standard practice in the international mining industry. They are meant to maximise the potential for environmentally sound and sustainable development by integrating environmental issues into development planning (Hunter et al. 1998; Jain et al. 1993; Knox 2002; Van Dyke 1993; Robinson 1992). Often included in an acceptable EIA model is a social impact assessment (SIA), which is defined as 'the process of assessing or estimating, in advance, the social consequences that are likely to follow from specific policy actions or project development' (Burdge and Vanclay 1996:1). SIAs are meant to cover 'all social and cultural consequences to human populations of any public or private actions that alter the ways in which people live, work, play, relate to one another, organize to meet their needs, and generally cope as members of society' (Burdge and Vanclay 1996:1). Furthermore, 'cultural impacts involve changes to the norms, values, and beliefs of individuals that guide and rationalize their cognition of themselves and their societies' (Burdge and Vanclay 1996:1). In the 1994 mining law, there are no specific measures calling for an EIA or SIA by the holder of the mining permit, let alone an independent third party, and there is no provision for public participation and public disclosure. To date, meaningful and thorough EIAs are not a part of the mining industry in Burma; they require a discursive democratic participation.

As Tun Myint (2003:292) has stated in relation to Burma, 'environmental governance is an inherently political process'. Many writers note that environmental permitting—meaning the decision to proceed with a project—is a political choice resulting from societal values and expectations (Joyce and Macfarlane 2001). By today's global standards, this political choice is meant to include local and possibly dissenting voices, which has been reflected, at least in part, through the implementation of fair and efficient EIAs and SIAs.

This is an area of concern for fair and effective environmental governance in Burma's formal mining sector: the local voice is not factored into development planning, let alone through fair and efficient impact assessments. If a governance regime does not fully consider the issues and interests of all layers—local, national and transnational— 'then that particular regime is less likely to achieve stated goals by means of fair and efficient governance process' (Myint 2002:109).

Furthermore, it is telling that there is effectively nothing legally explicit and precise in the mining laws beyond a statement of the requirement for a permit and detailing the rents and royalties that must be paid to the SPDC. Aside from direct percentage quotes (for example, copper calls for a 3–4 per cent royalty to be paid to the Ministry of Mines, the exact amount to be determined by the SPDC), there is little substantive or procedural content to the Myanmar Mining Law of 1994 that would constitute a basis for sound environmental governance. By its total exclusion of social and environmental considerations, and its sole emphasis on securing revenues, the Myanmar Mining Law's structure reveals its elemental purpose: economic governance, devoid of environmental and social measures.

The Monywa Copper Project

Part of the attraction to Burma for a company such as Ivanhoe Mines—aside from the plentiful mineral deposits—is the regulatory freedom with which the company can work, essentially having little to no responsibility to abide by domestic regulations (since none exist), and little to no procedural or corrective justice that might hold it accountable for any wrongs committed. The Mining Law of 1994 actually protects mining companies from liability, prosecution or fines (Gutter 2001:9). In virtue of these laws, and its location in Burma, Ivanhoe Mines' Monywa Copper Project is completely isolated from the outside world, and its activities are protected from scrutiny. A close consideration of the Monywa Copper Project example helps reveal the nexus formed by this type of irresponsible foreign investment, environmental degradation and Burma's economic-driven authoritarian governance.

The Monywa Copper Project is Burma's largest mine, located in central Burma, Sagaing Division, and comprises four sulfide-copper deposits named Sabetaung, Sabetaung South, Kyisintaung and Letpadaung. The first three (referred to collectively as S&K) are adjacent and because of that they have been developed as one project, while Letpadaung is approximately 10 kilometres southeast of S&K and is currently undeveloped.

The Monywa Copper Project is operated by the Myanmar Ivanhoe Copper Company Limited (MICCL), which is a company created by the 13-year-old 50–50 joint-venture agreement between Ivanhoe Mines and Number One Mining Enterprise (ME1). ME1 is a state-owned company and one of five companies that were created under the Ministry of Mines when General Ne Win seized power in 1962.[3] Ivanhoe Mines, the Canadian half of the joint venture, is listed on the Toronto and New York stock exchanges and the NASDAQ. The Monywa project started commercial production in 1999 with an annual production of 27,000 tonnes of copper cathode. The copper cathode produced at Monywa is certified with the London Metals Exchange, the global authority in copper trade, and it is sold to Marubeni Corporation, Japan's fifth largest trading company. Marubeni Corporation also financed the development of the Monywa mine through a US$90 million loan, which Ivanhoe Mines Limited repaid in full in August 2005 (Ivanhoe Mines Limited 2006a).

Methods and processes at Monywa

> The acid used at MICC is very strong and hurts my eyes, especially near the acid pond. Many of my friends work at the acid ponds and they told me how hard it is. They try to take care of the acid but it's not easy[4] (A miner at the Monywa Copper Project)

Copper extraction is notoriously messy and a potentially disastrous process anywhere, even when the most advanced methods are used (Ayres et al. 2002). At Monywa, the large-scale extraction of copper from the earth is an involved, technologically complex and multi-stage process. Though there can be some general similarities between disparate copper-mining operations throughout the world, 'there is no such thing as an "ordinary mining operation"' (Ripley et al. 1996:5). Mineral production and processing depends on a number of factors, such as terrain, geology and mineralogy, and is always unique from site to site.

MICCL uses an advanced process called the solvent extraction-electro winning (SX-EW) method. In basic terms, copper ore is mined, crushed and stacked on an allegedly 'impermeable' liner. The ore is then sprayed with a leaching solution containing sulphuric acid, which dissolves the copper. The resulting copper-rich solution is then treated with an organic solvent and an electrical current, with an end product of 45-kilogram sheets of 99.999 per cent pure copper. The process creates a toxic waste referred to as tailings, and these tailings, through a process known as acid mine drainage (AMD) or acid rock drainage (ARD), can contribute to environmental degradation of rivers, ground water and soil. The tailings can also find their way into the hands of locals, who attempt to extract the little copper that remains in the waste by using an environmentally destructive process of artisanal mining called *dohtar* in Burmese.

Monywa is an open-pit mine, which is the first choice of mining companies as this is the most economical form of mining. They are also the most devastating to the landscape. Open-pit mines account for the largest human-made holes on the planet, holes that can impact only irreversibly on the ecosystems they displace.

Copper ore contains only a small percentage of the desired metal, so vast amounts of it must be dug up for a comparatively small amount of copper. Industry standards currently hold that mining copper ore is worthwhile—that is, profitable—when there are at least 2kg of copper for every 1,000kg of ore, but of course this standard fluctuates with technological advancement, so that as technological efficiency improves it will theoretically become progressively easier to extract smaller percentages of copper to ore.

Leaching, as mentioned above, is a chemical process that involves administering toxic acid over the heaps of ore, which dissolves the metal out of the ore. At Monywa, sulphuric acid is the active acid in the leaching solution. The leach pads on which the giant heaps of ore sit are commonly referred to as impermeable, suggesting, misleadingly, that the toxic chemicals administered to the ore are safely trapped. Heap leach pads are, however, tragically prone to toxic leaks and spills—more likely without fair and efficient environmental governance.

Furthermore, the SX-EW process is theoretically meant to recycle the sulphuric acid, so that the acid administered in the SX process is recovered in the EW process, and is used again. Additional acid, however, is required of the process, as demonstrated in part by MICCL's importing of sulphuric acid and plans to construct an on-site acid factory. Side effects of this process pose considerable threats to surface and ground water. Copper-mining experts commissioned in part by the World Business Council for Sustainable Development refer to the process as follows

> The heap leaching (SX) process is another source of waste, seldom discussed. Large amounts of sulphuric acid (a smelter by-product) are used for this purpose…In principle the sulfur in the system is simply recycled as sulfuric acid and returned to the leaching operation. On

the other hand…it seems clear that much more sulfuric acid is used for leaching than is recovered at the EW stage. The residue presumably reacts with other materials in the ore or concentrate. Since most sulfates are somewhat soluble, they presumably find their way into surface waters or ground water. The literature does not discuss this point. (Ayres et al. 2002:28)

Lastly, the method used at Monywa requires a large amount of energy and water, which is an environmental and economic concern anywhere, let alone in a country such as Burma with weak infrastructure and poor resource management.

Environmental management system

Ivanhoe Mines Limited boasts that MICCL operates according to international standards, including ISO 14001, which is an environmental management system (EMS) developed by a Geneva-based non-governmental organisation (NGO). The inherent weakness of this EMS is that it is voluntary, and the extent of the application of ISO 14001 is decided entirely by the company itself. That is, companies design product and process-specific standards and establish their own objectives and goals to achieve these standards.

While the growth of international standards throughout the world has some observers enthusiastic, local people in mining communities world-wide fail to see or experience noteworthy social and environmental improvements from the management system. Despite Ivanhoe Mines Limited boasting of MICCL's environmental integrity and social stewardship (Ivanhoe Mines 2004a), a less enthusiastic local voice can be found at Monywa. As one miner at the MICCL mine notes, 'The forest and trees are gone from the area where MICCL is located. Trees are unable to grow in this area any more. I think it is because there is a lot of acid in the soil surrounding the mine site.'[5]

It has also been argued that voluntary agreements such as ISO 14001 can actually complicate rather than facilitate governance. Bruce Paton surveyed empirical and political economy studies of negotiated, voluntary, regulatory frameworks and found that 'negotiated agreements…often reduce transparency—the ability of outside parties to observe both the process and the outcomes of a policy—relative to regulations'. He found that 'empirical studies document that negotiated agreements have permitted significantly less community and non-governmental organization participants than previous regulatory policies' and that 'the political economy studies argue that both industry and regulatory agencies have favored voluntary approaches precisely because they reduce the influence of both legislative bodies and environmental groups on policy outcomes' (Paton 1999:1, 26).

This corresponds with Ivanhoe Mines' recent statement that NGOs and local people were a nuisance to mining, and were to be avoided. Robert Friedland, founder and Executive Chairman of Ivanhoe Mines Limited, recently noted at a Global Resources conference that 'you want [your mine] to be near the market, but you don't want people around your mine because people near your mining project are a real nightmare'. In reference to Ivanhoe Mines' Oyu Tolgoi Copper Project in the Gobi Desert in Mongolia, he explained that 'the nice thing about this [is], there's no people around, the land is flat, there's no tropical jungle, there's [*sic*] no NGOs'. Continuing, quite candidly, he added that 'the nice thing about the Gobi is, there's no railroad tracks in the way, there are no people in the way, there are no houses in the way' (Friedland 2005). This rather crude attitude communicates the view that local people and NGOs are a hindrance and obstacle rather than a critical component and equal partner to fair and efficient environmental governance.

Recent developments

Ivanhoe Mines Limited continues to seek international finance for a US$400 million expansion of the Monywa Copper Project, which would potentially double the size of the Monywa operation (*The Irrawaddy* 2005b). This involves the development of the Letpadaung ore deposit, which will make Monywa one of the largest copper mines in Asia in terms of recovery rates and annual production.

The company began 'reviewing strategic alternatives' to its investment in Monywa as early as March 2004 (Ivanhoe Mines 2004b). At the time of writing, the company was seeking to sell a 25 per cent stake in the Monywa Copper Project to a South Korean consortium that included Daewoo International, a company with an already large and controversial investment in vast natural gas deposits off the Arakan (Rakhine) coast in western Burma, referred to as the Shwe gas project.[7] The rest of the possible South Korean consortium includes Korean Resources Corporation and Taihan Electric Wire. A memorandum of understanding was purportedly signed in January 2006 and the official deal was supposed to be finalised in July 2006, pending a due-diligence project analysis (*The Korean Herald* 2006).

As a joint partner with the military through the Myanmar Ministry of Mines, Ivanhoe Mines Limited is obligated to keep an amicable relationship with the SPDC if it expects to continue the business relationship. Ivanhoe Mines Executive Chairman, Robert Friedland, has empathised with the regime, attesting to its integrity by saying, 'they really love their country' (Moody 1999:52). He went on to explain Ivanhoe Mines' ethical bottom line, stating that if the military started 'killing students en masse, we would have to re-evaluate our involvement in Myanmar' (Moody 1999).

With an ultimate concern for keeping copper production high and production costs low, the MICCL was relatively successful in Burma, with an operating profit from the first nine months of 2005 of a reported US$29.1 million, a 62 per cent increase from the previous year. Ivanhoe Mines also reported a 52 per cent increase in revenue for MICCL, due mostly to the fluctuating price of copper (Ivanhoe Mines Limited 2005).

Recently, however, Ivanhoe Mines reported difficulties with the Monywa Copper Project. There are at least four reasons for this. First, the company's insurance broker and offshore bank terminated its relationship with MICCL (Ivanhoe Mines Limited 2006c). Ivanhoe Mines claims this termination was a result of sanctions imposed on

Burma by the US government, but it would not have been sanctions *per se* preventing the business relationship from continuing. Regardless, this caused the mine site to close down for March 2006 because of an inability to cover basic operating costs and related insurance risks. Second, the company reported a steady decrease in mine production during 2005 and 2006, due in part to a drop in copper grades at the mine site (Ivanhoe Mines Limited 2006c). Second-quarter results released by the company in August 2006 reflected these problems. While production restarted on 2 April 2006, the company faced a loss of 44 per cent compared with the same quarter in 2005 (Ivanhoe Mines Limited 2006d).

Third, making the situation even more difficult for Ivanhoe Mines, the military regime has refused to issue import permits for much needed mining equipment (Ivanhoe Mines Limited 2006a). Without this equipment, the S&K mine cannot be developed, posing a significant obstacle to current and future profitability. Ivanhoe Mines has acknowledged that this 'could result in significant decreases in copper production for 2006 and subsequent years' (Ivanhoe Mines Limited 2006a). Fourth, since 2005, Ivanhoe Mines has been engaged in a dispute with Burma's tax authorities, which imposed an 8 per cent tax on all export sales (Ivanhoe Mines Limited 2006a). This tax was imposed retroactively from 1 January 2003 and is understandably opposed by the company, which has filed a complaint. Ivanhoe Mines estimates this will cost it approximately US$11 million (Ivanhoe Mines Limited 2006a).

This is all particularly difficult for the company considering that the price of copper on the London Metals Exchange recently skyrocketed to nearly double the price it was in December 2005 (UNIRIN 2006). Furthermore, the difficult negotiations recently undertaken with the government do not bode well for the expansion plans or the sale of its interests to the Korean consortium. An official deal to be signed with the Koreans in July 2006 had not materialised, leading some to speculate that the investment environment had soured considerably (McClearn 2006).

Artisanal mining at Monywa

Artisanal mining or *dohtar* drives a cycle of poverty and environmental degradation, and in Burma there is, to date, no effective environmental governance to address this widespread practice. At Monywa, individuals enter into artisanal mining because of poverty and sometimes because pre-existing environmental degradation rendered subsistence farming difficult or impossible. In turn, the practice perpetuates the poverty and environmental degradation that ushered them to artisanal mining in the first place.

Local people at Monywa have complained that they can no longer farm their land due to high levels of sulphuric acid in the soil,[7] pushing some to artisanal mining, which only adds to the degradation that adversely impacted them in the first place.[8] This phenomenon has created a local economic shift—which has occurred elsewhere in Burma—from a subsistence-based to a cash-based economy. This shift increases inflation, which increases the difficulty of purchasing goods for survival, such as cooking oil, fuel, clean water, medicine and so on (EarthRights International 2004). Informal artisanal mining is not economically sustainable, requiring no capital inputs and thus no added return on inputs; it is a vehicle for the perpetuation of poverty.

Artisanal mining has been enabled by the larger Monywa Copper Project operation. It is fairly simple and highly toxic, involving the manual extraction of small amounts of copper from the tailings waste of the Monywa mine site. In some cases, the waste is carried manually from the larger mine site or its immediate vicinity and placed in small pools of water. Sulphur is added, then the mixture is boiled. Next, tin milk cans are added, causing a chemical reaction, and the resulting acid slowly dissolves the cans. The process takes approximately 10 days and, when complete, leaves copper ore in a highly toxic pool of water. The copper is removed by hand with little or no safety precautions, and sold to local and Chinese businessmen. There is no clean-up. Since 2005, EarthRights International has recorded the deaths of at least three artisanal miners on the property of the Monywa Copper Project. To date, there has been no real response and no proposed environmental governance plan to address the widespread artisanal mining, despite the fact that the practice occurs in plain view and in the immediate vicinity of the MICCL operation.

The ability of miners and locals to effect change or voice concerns about practices at Monywa is limited. The security situation at the copper project is unique but, in effect, no different than the standard practices in Burma. Ivanhoe Mines has claimed publicly that it did not make arrangements to have the mine site at Monywa secured by the *Tatmadaw* (military).[9] One local miner told EarthRights International that there were 'military personnel to maintain security for the mine site. But they do not wear uniforms so it is difficult to tell who they are around the mine site.' Some miners are unsurprisingly hesitant to speak honestly within the limited privacy of their own living quarters: 'I don't want to complain too much about the government because the walls in my room are thin and many people living next to me can hear me right now. I do not want them to report my words to the government'.[10]

The precautionary principle and a shift in power

Highlighting weaknesses and problems with environmental governance and mining in Burma begs questions of how the situation could improve. What would fair and effective environmental governance of mining in Burma look like? How could advances be achieved? These questions are immense in scope and beyond the limits of this chapter, but it is worth highlighting a few key areas relevant to a positive future of environmental governance of mining in Burma.

This section discusses the 'precautionary principle' and the emerging requirements of companies for responsible mining, highlighting specific areas in need of improvement at Monywa. This is followed by a section on local participation in environmental governance at Monywa, which highlights Burma's unique tradition of relying on 'respected insiders' for conflict resolution.

The precautionary principle is a well-recognised principle of international law codified in the Rio Declaration (UNEP 1992) and other international environmental instruments.[11] The principle requires states to take proactive precautionary measures 'where there are threats of serious or irreversible [environmental] damage' (Principle 15, Rio Declaration on Environment and Development). This applies likewise to private corporate actors, and some such measures are expected to be undertaken by an objective third party, for example, as with an EIA.

Scholars have noted that the real gravity of the precautionary principle is that it has shifted the burden of proof from local communities and NGOs, for example, to companies and governments undertaking the practice in question. In effect, it is what Robert Durant has referred to as 'the obverse of traditional regulatory approaches', traditional approaches that naively assumed safety until proven harmful (Durant and Boodphetcharat 2004:105). There is a certain corporate distaste for this shift. That is, mining companies do not appreciate the added responsibility that comes with the burden of proof regarding environmental safety, and generally they are quite forthright about that sentiment, as it is an expensive burden indeed.

Despite that, this shift has emerged simultaneously with what EarthRights International refers to as earth rights—namely, environmental and human rights. In short, the shift reflects the nascent, emerging environmental duties of corporate actors, while simultaneously reflecting the emerging conceptualisations of environmental and human rights (Center for Economic and Social Rights 1994), legal mechanisms to uphold those rights, and more frequent and widespread empowerment of local communities. From a broad perspective, the precautionary principle represents a somewhat revolutionary shift in power, albeit a revolution unrealised and still in motion. Alas, in this context, many local communities truly have nothing to lose.

At Monywa, the emergence of fair and effective environmental governance would reflect that shift in power in numerous ways. It would involve objective third-party impact assessments and environmental monitoring, as well as measures aimed to enfranchise the local population in the process of environmental governance. Ideally, the mine would be opened up to at least minimal public scrutiny, including transparency in payments. Regarding the latter, the Extractive Industries Transparency Initiative (EITI) calls on governments to disclose how payments from extractive industries are distributed for national and regional priorities, and they are becoming increasingly successful in achieving greater financial transparency, and thus better governance (see www.eitransparency.org). There has been no reporting of the amounts paid to the SPDC or where that money has been spent.

Fair and effective environmental governance will also include continuing community environmental education initiatives, and culturally relevant measures for environmental conflict resolution, including culturally relevant interpretations of citizen-based and group-based participatory processes (see below). This will provide mechanisms for corporate accountability, it will empower local actors and ideally offer redress when appropriate.

Ivanhoe Mines Limited[12] and MICCL should also go beyond their current environmental and social reporting practices, which can be found on Ivanhoe Mines' web site, where the most recent available health and environmental report is from 2004 (Ivanhoe Mines Limited 2004a). The glaring inadequacy of Ivanhoe Mines' current reporting on achieving safety, health and environmental goals at Monywa is that it is completely unverifiable. The company should therefore participate in reporting through specific and measurable indicators that can be verified independently, as recommended by the Global Reporting Initiative's *Sustainability Reporting Guidelines* (GRI 2002). As mentioned previously, ISO 14001, which Ivanhoe employs, is an inadequate system for environmental monitoring.

It is widely regarded that artisanal and small-scale mining (ASM) holds considerable potential for reducing poverty in countries such as Burma (see 'Artisanal Mining for Sustainable Livelihoods' 1999). There is a certain corporate distaste for this, demonstrated by Ivanhoe Mines' repeatedly flat references to artisanal miners as 'illegal miners' and the company's failure to facilitate responsible artisanal mining in the area.

A clear environmental governance plan should be developed to engage Monywa's artisanal mining community in a way that will benefit that community—a plan that involves more than simply clearing out the artisanal miners.

Conclusion: local participation and respected insiders

If there is one certainty of fair and effective local participation in environmental governance, it is that there is no universal monolithic system of rules, regulations and processes simply awaiting implementation and practice. Just as disparate copper-mining operations can differ vastly, so too do local potentialities for environmental governance participation (Medowcroft 2004; and, for a contrasting account, Leone and Giannini 2005). There are, however, two consistent features of effective local participation in environmental governance: it must involve local people and have, to some degree, cooperation and support from relevant institutions and stakeholders. That is, it's a multi-stakeholder affair, and moreover one that presupposes the recognition of the right to organise.

Environmental conflict resolution is a tool for recourse and 'for building common purpose' between stakeholders (O'Leary et al. 2004:324). Scholars note the importance of understanding the many varieties of environmental conflict resolution interventions 'as complex systems embedded in even larger complex systems' (O'Leary et al. 2004:324). In other words, the wider spatial, temporal, economic, social, cultural and political contexts of the specific environmental conflict resolution are relevant for building common purpose between stakeholders. In Burma, conflict resolution is undertaken quite differently from dominant Western models. EarthRights International conducted research for five years on traditional methods of conflict resolution and its relationship to resource-based conflict at the local level in Burma. That research resulted in *Traditions of Conflict Resolution in Burma* (Leone and Giannini 2005), which argues that conflict resolution in Burma is based more on interpersonal respect and a tradition of local 'respected insiders' than on assumptions of the objectivity of 'third-party outsiders'. Whereas official administrative and court-based proceedings provide a level of comfort and trust to the Western sensibility, these are the very institutions and processes that might cause local villagers in Burma to feel uncomfortable and distrustful. The report contends that 'the prospects for peace and earth rights protection' hinge on this respected insider model, adding that such respected insider 'practices may serve as models for community-based natural resource management' (Leone and Giannini 2005:1–2). Effective local participation in environmental governance in Burma will necessarily involve a unique tradition-based paradigm developed by local Burmese themselves.

While third-party outsiders are less likely to gain genuine traction in communities in Burma, this is not meant to undermine the need for objective third-party EIAs and environmental monitoring at large-scale mining operations such as Monywa. Rather, it simply indicates the unique needs that must be considered for fair and effective local participation in environmental governance of mining in Burma. While administrative and judicial proceedings can make the average Burmese villager uncomfortable, the same cannot be said for the rule of law and justice (which are largely absent in Burma), which will be accepted wholly by the average Burmese, particularly by those whose human rights have been violated.

As Tun Myint (2003) has suggested, the successes and failures of environmental governance are determined largely by how natural resources are used and managed at the local level. This chapter approached a genuine inquiry into the state of environmental governance of mining in Burma motivated by a genuine concern for the natural environment and the people of Burma who depend on it. It interpreted current environmental governance of mining natural resources in Burma as largely inadequate, weak and ostensibly favourable to corporate interests over the public interest and the natural environment. Burma's economic, social, cultural, political and environmental future depends on changing this.

Notes

1 In 1962, separate companies were created to handle specific minerals, including investment in those minerals and concessions granted. The companies are Number One Myanmar Enterprise (ME1), Number Two Myanmar Enterprise (ME2), Number Three Myanmar Enterprise (ME3), Myanmar Gems Enterprise (MGE), Myanmar Pearl Enterprise (MPE) and Myanmar Salt and Marine Chemical Enterprise. See http://mining.com.mn/Mines/pltim.asp

2 Though concessions are awarded expressly by the government, Northern Star, a Chinese mining company, has control of all concessions granted in Kachin State, reflecting its close relationship with the SPDC and a general shift in power over concessions.

3 See Note 2.

4 EarthRights International interview No.038, on file with author.

5 EarthRights International interview No.038, on file with author.

6 Daewoo International is the largest stakeholder in the Shwe gas project, which involves the exploration and development of vast natural gas deposits worth upwards of US$80 billion, located off the Arakan coast in western Myanmar. This mega-development project will adversely affect more people than any other project in Myanmar's history, and will be the SPDC's largest source of revenue, generating up to US$17 billion in the course of 30 years, according to *Supply and Command*, a recent report by the Shwe Gas Movement. Available from http://shwe.org

7 EarthRights Internationl interviews, on file with author.

8 EarthRights Internationl interviews, on file with author

9 Letter available at http://www.amnesty.ca

10 EarthRights Internationl interview No.038, on file with author.

11 See, for example, the European Union's Registration, Evaluation, and Authorisation of Chemicals (2003), the Cartagena Protocol on Biosafety (2003) and the Stockholm Convention on Persistent Organic Pollutants (2004). The principle specific to environmental legislation is also appearing in national legislation in, for example, France and the United States (San Francisco, CA) cites the precautionary principle in Article one of the city's 2003 environmental legislation. Thanks to Carl Byers for clarification on this point.

12 On March 30, 2007 Ivanhoe Mines reported that due to requirements of its partnership with Rio Tinto in Mongolia, the company transferred all of its assets in Burma to an independent trust. This means the company does not have control over the sale of the assets, but until the assets are sold the company continues to collect revenues and continues to operate the mine in partnership with the Ministry of Mines and the military regime in Burma.

References

ABARE and Mekong Economics, 2005. *Enhancing ASEAN Minerals Trade and Investment: country reports*, December. Available from http://www.aadcp-repsf.rog/docs/04-009b-FinalCountryReport.pdf.

Ayres, R.U., Ayres, L.W. and Rade, I., 2002. *The Life Cycle of Copper, Its Co-Products and By-products*, International Institute for Environment and Development, London. Available from http://www.iied.org/mmsd/mmsd_pdfs/ayres_lca_main.pdf.

Burdge, R.J. and Vanclay, F., 1996. 'Social impact assessment: a contribution to the state of the art series', *Impact Assessment*, 14(1):59–86.

Center for Economic and Social Rights, 1994. *Draft Declaration of Principles on Human Rights and the Environment, E/CN.4/Sub.2/1994/9, Annex I (1994)*, (referred to as the 'Ksentini Principles'), CEDHA, New York. Available from http://www.cedha.org.ar/en/documents/25_documents.

Downing, T.E., 2002. *Avoiding New Poverty: mining-induced displacement and resettlement*, No. 52, April, International Institute for Environment and Development, London. Available from http://www.iied.org/mmsd/mmsd_pdfs/058_downing.pdf.

Durant, R. and Boodphetcharat, T., 2004. 'The precautionary principle', in R. Durant, D.J. Fiorino and R. O'Leary (eds), *Environmental Governance Reconsidered: challenges, choices, and opportunities*, MIT Press, Cambridge, Mass.:105–44.

EarthRights International, 2004. *Mining, Gender, and the Environment in Burma*, EarthRights International, Chiang Mai. Available from http://www.earthrights.org/burmareports/mining_gender_and_the_environment_in_burma.html

EarthRights International and Karen Environmental and Social Action Network, 2003. *Capitalizing on Conflict*, October, EarthRights International, Chiang Mai. Available from http://www.earthrights.org/files/Reports/capitalizing.pdf

Energy Planning Department, Myanmar Ministry of Energy, 2001. *Myanmar Foreign Investment Law*. Available from http://www.energy.gov.mm/Incentive_1.htm

Friedland, R., 2005. Nothing like it on planet earth—Robert Friedland's Tour d' Tolgoi, Address delivered at the BMO Nesbitt Burns 2005

Global Resources Conference, 7 March, Tampa, Florida. Available from http://www.resourceinvestor.com/pebble.asp?relid=9010

Global Reporting Initiative, 2002. *Sustainability Reporting Guidelines*, updated 2006, Global Reporting Initiative. Available from http://www.globalreporting.org/guidelines/2002.asp

Gutter, P., 2001. 'Environment and law in Burma', *Legal Issues on Burma*, 9 (August). Available from http://www.ibiblio.org/obl/docs/LIOB09-environment_and_law_in_burma.htm

Hentschel, T., Hruschka, F. and Priester, M., 2003. *Artisanal and Small Scale Mining: challenges and opportunities*, International Institute for Environment and Development, London. Available from http://www.iied.org/mmsd/index.html

Hunter, D., Salzman, J. and Zaelke, D., 1998. *International Environmental Law and Policy*, Foundation Press, New York.

Images Asia and Pan Kachin Development Society, 2004. *At What Price? Gold mining in Kachin State, Burma*, Nopburee Press, Chiang Mai. Available from http://www.ibiblio.org/obl/docs/gold%20pdf1.pdf.

International Institute for Environment and Development, 2002. 'Artisanal and small scale mining', *Breaking New Ground: mining, minerals and sustainable development*, Earthscan Publications, London:313–34. Available from http://www.iied.org/mmsd/index.html.

Ivanhoe Mines Limited, 2004a. *MICCL Monywa Copper Project: safety, health, and environment report (2004)*. Available from http://ww.ivanhoe-mines.com.

——, 2004b. *MICCL Monywa Copper Project: safety, health and environmental report 2004*. Available from http://www.ivanhoe-mines.com/s/MonywaCopper.asp.

——, 2005. 'Ivanhoe Mines announces Q3, including an operating profit of $9.4 million', *Quarterly Report*, 10 November. Available from http://ww.ivanhoe-mines.com.

——, 2006a. *Annual Information Form for the Year Ended December 31, 2005*, 30 March 2006. Available from http://www.sedar.com.

——, 2006b. *Fact File: the Monywa Copper Project*. Available from http://ww.ivanhoe-mines.com.

——, 2006c. *Quarterly Financial Report for the Three and Six Months ended June 30, 2006*. Available from http://www.sedar.com.

——, 2006d. Ivanhoe announces Q2 2006 results, News release, 14 August. Available from http://ww.ivanhoe-mines.com.

Jain, R.K., Urban, L.V., Stacey, G.S. and Balbach, H.E., 1993. *Environmental Assessment*, McGraw Hill, Austin.

Joyce, S.A. and Macfarlane, M., 2001. *Social Impact Assessment in the Mining Industry: current situation and future directions*, International Institute for Environment and Development, London. Available from http://www.iied.org/mmsd/mmsd_pdfs/social_impact_assessment.pdf.

Knox, J., 2002. 'The myth and reality of transboundary environmental impact assessment', *American Journal of International Law*, 96(2):291–319.

Leone, F. and Giannini, T., 2005. *Traditions of Conflict Resolution in Burma: respected insiders, resource-based conflict and authoritarian rule*, EarthRights International, Chiang Mai. Available from:http://www.earthrights.org/files/Reports/ctwp_paper.pdf

MacLean, K., 2004. 'Extractive industries in Burma: mining in comparative context', in EarthRights International (ed.), *Mining, Gender, and the Environment in Burma*:2. Available from http://www.earthrights.org.

McClearn, M., 2006. 'Mine games: Ivanhoe Mines', *Canadian Business*, 22 May–4 June. Available from http://www.canadianbusiness.com.

Medowcroft, J., 2004. 'Deliberative democracy', in R. Durant, D.J. Fiorino and R. O'Leary (eds), *Environmental Governance Reconsidered: challenges, choices, and opportunities*, MIT Press, Cambridge, Mass.:183–218.

Mélainé, J., Kim, M., Hester, S., Berry, P., Ball, A. and Schneider, K., 2005. *Enhancing ASEAN Minerals Trade and Investment: country reports*, December, Regional Economic Policy Support Facility/ASEAN–Australia Development Cooperation Program, Jakarta. Available from http://www/aadcp-repsf.org/docs/04-009b-FinalCountryReport.pdf

Miranda, M., Chambers, D. and Coumans, C., 2005. *Framework for Responsible Mining: a guide to evolving standards*, Center for Science in Public Participation, Bozeman, Montana. Available from http://www.frameworkforresponsiblemining.org.

Moody, R., 1999. *Grave Diggers: a report on mining in Burma*, Mining Watch Canada, Ottawa. Available from http://www/miningwatch.ca/updir/Grave_Diggers.pdf.

Myanmar Ministry of Mines, 2004. *Government Policy and Legislation on Investment in Minerals*. Available from http://www.mining.com/mm/Mines/pltim.asp.

Myint, T., 2002. 'Managing complexities in global environmental governance: issues-interests-actors network model for the transnational environmental governance in the Mekong River Commission and the International Commission for the Protection of the Rhine', in F. Bierman, R. Brohm and K. Dingwerth (eds), *Global Environmental Change and the Nation State: proceedings of the 2001 Berlin conference on the human dimensions of global environmental change,* Potsdam Institute for Climate Impact Research, Potsdam:107–17.

——, 2003. 'Democracy in global environmental governance: issues, interests, and actors in the Mekong and the Rhine', *Indiana Journal of Global Legal Studies,* 10(1):287–314.

O'Leary, R., Nabatchi, T. and Bingham, L.B., 2004. 'Environmental conflict resolution', in R. Durant, D.J. Fiorino and R. O'Leary (eds), *Environmental Governance Reconsidered: challenges, choices, and opportunities,* MIT Press, Cambridge, Mass.:323–54.

Paton, B., 1999. 'Voluntary environmental initiatives and sustainable industry', *1999 Greening of Industry Network Conference—Best Paper Proceedings,* Greening of Industry Network. Available from http://www/greeningofindustry.org.gin1999/Paton.pdf.

Ripley, E.A., Redman, R.E. and Crowder, E.A., 1996. *The Environmental Effects of Mining,* St Lucie Press, Delray Beach.

Robinson, N.A., 1992. 'International Trends in Environmental Impact Assessment', *Boston College Environmental Affairs Law Review,* 19(3):591–610.

Rondinelli, D. and Vastag, G., 2000. 'Panacea, common sense, or just a label?: the value of ISO 14001 environmental management systems', *European Management Journal,* 18(5):499–510.

Shwe Gas Movement. *Supply and Command,* report by the Shwe Gas Movemnent. Available from http://shwe.org

State Law and Order Restoration Council, 1988. *The Myanmar Foreign Investment Law, SLORC Law 10/88,* 30 November. Available from Burma Lawyers Council: http://www/blc-burma.org/html/Myanmar%20Law/lr_e_ml88_10.html.

——, 1994. *Myanmar Mines Law 1994, Chapter V, Article 15.* Available from Burma Lawyers Council http://www/blc-burma.org/html/Myanmar%20Law/lr_e_ml194_08.html.

Tan, A.K.J., 1998. *Preliminary Assessment of Myanmar's Environmental Law*, Faculty of Law, National University of Singapore. Available from http://www.law.nus.edu.sg/apcel/dbase/myanmar/reportmy.html#nep.

The Burma Campaign (UK), 2004. *The EU and Burma: the case for targeted sanctions*, Burmacampaign.org.uk, London. Available from http://www.burmacampaign.org.uk/reports/targeted_sanctions.htm.

The Irrawaddy, 2005a. 'Foreign investment in Burma hits US$7.6 billion', *The Irrawaddy Online Newsletter*, 18 November.

——, 2005b. 'Ivanhoe going for gold', *The Irrawaddy Online Newsletter*, 20 November. Available from http://www.irriwaddy.org/aviewer.asp?a=738&z=109.

The Korea Herald, 2006. 'Korean firms to mine copper in Myanmar', *The Korea Herald*, 13 April.

United Nations Development Programme, 1972. *Declaration of the United Nations Conference on the Human Environment (1972)*, (referred to as 'The Stockholm Declaration'), United Nations Development Programme, Stockholm. Available from http://www.unep.org.

——, 1999. *Artisanal Mining for Sustainable Livelihoods*, United Nations Development Programme, New York. Available from http://www.undp.org.

United Nations Environment Program, 1992. *Rio Declaration on Environment and Development (1992)*, United Nations Environment Program, Rio de Janeiro. Available from http://www.unep.org.

United Nations Integrated Regional Information Networks, 2006. *Zambia: record copper prices, but mine region yet to benefit*, July, United Nations Office for the Coordination of Humanitarian Affairs. Available from http://www.irinnews.org.

van Dyke, J.M., 1993. 'Sea shipment of Japanese plutonium under international law', *Ocean Development and International Law*, 24(4):399–430.

Acknowledgments

The author would like to thank Ann Putnam, Gideon Lundholm, Naing Htoo, Chana Maung, Carol Ransley and Alisa Loveman for their various levels of support. Additional thanks to Tyler Giannini, Thom Ringer, Tun Myint, Ken MacLean and Nicholas Pharris-Ciurej for their helpful advice on earlier drafts.

11 Spaces of extraction
Governance along the riverine networks of Nyaunglebin District

Ken MacLean

Contemporary maps prepared by the State Peace and Development Council (SPDC) place most of Nyaunglebin District in eastern Pegu Division. Maps drawn by the Karen National Union (KNU), however, place much of the same region within the western edge of *Kaw Thoo Lei*, its term for the 'free state' the organisation has struggled since 1948 to create. Not surprisingly, the district's three townships have different names and overlapping geographic boundaries and administrative structures, particularly in remote regions of the district where the SPDC and the KNU continue to exercise some control. These competing efforts to assert control over the same space are symptomatic of a broader concern that is the focus here, namely: how do conflict zones become places that can be governed? What strategies and techniques are used to produce authority and what do they reveal about existing forms of governance in Burma? In considering these questions, this chapter explores the emergence of governable spaces in Shwegyin Township, which comprises the southern third of Nyaunglebin District (Figure 11.1).

Figure 11.1 Shwegyin Township Mining Area

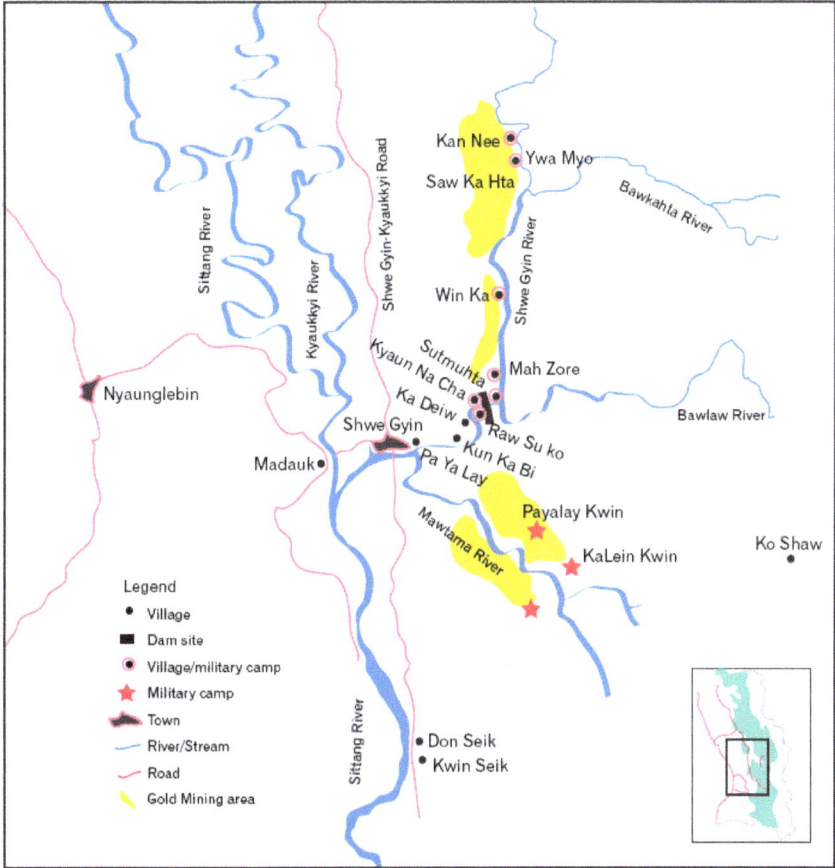

For decades, the SPDC and the KNU fought one another to control the riverine systems in Nyaunglebin District, and with them the flow of people, information, food and other commodities in the region. During the mid 1990s, efforts to extract the district's natural resources intensified and more regulated forms of violence have since largely replaced lethal ones, at least where primary commodities are found. This chapter describes how the topography of these riverine systems shapes struggles by different state and non-state actors to control access to such valuable resources. Special attention is focused on the different ways military battalions and private business interests compete and collude with one another to produce a compliant labour force in low-lying areas where alluvial gold deposits are extracted and hydroelectricity is to be utilised. The data, drawn from field research conducted in the township between 2001 and 2005, reveal some of the contradictions that have accompanied efforts to consolidate centralised state control of Shwegyin Township (MacLean and Mahn Nay Myo 2002; ERI and KESAN 2003; ERI 2005; MacLean forthcoming).

Governable spaces

The term 'governable spaces' was coined by Nikolas Rose to explore the nature and the practice of government—what Foucault defined in his path-breaking lectures on governmentality as the 'conduct of conduct' (Rose 1999:31–4; Foucault 1991 [1978]). Who, in other words, can govern? What constitutes governing in a particular cultural and historical context? And what or who is being governed? To help answer these questions, which are far more complex than they first appear, Rose disaggregates the practice of government into four components so as to more fully examine how changes in the relationship between resources, territoriality and identity produce distinct kinds of spaces.[1] The first component concerns the ways by which sub-populations are defined *vis-à-vis* one another along axes of difference such as age, gender, language, ethnicity, religion, class, comportment and so on. The second involves the techniques of government through which authority is constituted and rule accomplished. The third emphasises the types of thought, calculation and expertise brought to bear in government—concerns that are not reducible either to those of ideology or economics. Lastly, the fourth stresses the forms of identification that produce governable subjects, who need specific kinds of 'official' documentation to be eligible for services and to exercise particular rights (see Watts 2004). All four components are, of course, relevant for understanding the diverse forms of government that have emerged in Burma's former conflict zones. Moreover, the components are interrelated and, ideally, they should be examined in relation to the others rather than separately. The discussion that follows, however, focuses more narrowly and concretely on the second component in order to make sense of the socioeconomic and ecological transformations that have occurred in Shwegyin Township in the past decade (c. 1996–2006).

It is important to note at the outset that these transformations, which will be described shortly, have resulted in contradictory outcomes, which simultaneously extend and fragment centralised state control of spaces where natural resources—in this instance, gold and hydroelectricity—are located.[2] As a consequence, current politico-administrative maps of Burma have become little more than 'cartographic illusions' (Ohmae 1995:7) that fail to depict accurately how these areas and the resource concessions located within them are governed—and by whom. A more accurate representation would require dynamic maps with multiple layers to convey how momentary economic alliances anchored to particular resource bases have created 'mosaics of territorial control' (Hardin 2002:ii), which have further dislocated populations and devastated the ecosystems they rely on. The findings also complicate analyses that continue to stress ethno-nationalist struggles as the primary cause of the SPDC's counter-insurgency campaigns, something it regularly justifies by claiming that the *Tatmadaw* (army) is the only institution capable of preserving 'national unity' and the 'territorial integrity' of Burma (ICG 2001; Houtman 1999:59–120).

The evidence presented here instead indicates that revenue streams and not national security concerns have long dominated the agenda of the *Tatmadaw*'s battalions in Nyaunglebin District. Moreover, regulatory agencies there have little or no ability to enforce existing laws related to the sustainable management of the country's resources. Recent reports published on the extractive industries operating elsewhere in Burma suggest this state of affairs is not limited to the district's waterways, but increasingly effects the country as a whole (see, for example, Global Witness 2003; Images Asia and PKDS 2004; Karen Rivers Watch 2004; and Thaung and Smith in this volume). If true, this means that much of Burma's remaining natural capital is under serious threat.

Background

Nyaunglebin District is a study in contrasts. The *Tatmadaw* has controlled large areas of the district for decades, especially urban areas and the low-lying plains, which form a narrow strip of fertile land on both sides of the Sittaung River. As a consequence, before 2006, these areas experienced only three waves of large-scale, state-sponsored violence, each of which entailed the forced relocation of thousands of Karen civilians: in 1975–1982, 1988–1990 and 1997–99 (Karen Rivers Watch 2004:26; BERG 1998). Interestingly, oral histories collected from those affected indicate that the extractive industries continued to operate through all but the worst moments of violence during each of these waves (MacLean forthcoming; ERI and KESAN 2003:27). Official statistics also support these claims.

Despite the violence, mining operations in Shwegyin Township were able to extract significant, if varying, amounts of gold during the third wave: 31 kilograms in 1988; 124kg in 1989; 26kg in 1990; and 145kg in 1991 (Pui-Kwan 1991:62). While further research is necessary, the information suggests counter-insurgency operations were not antithetical to the pursuit of profit; rather, they appear to have facilitated the expansion and intensification of the extractive industries in the township (see also Frynas 1998; Ferguson 2006).[3] This is not to suggest that violence has disappeared entirely; abuses remain widespread and continue to grow more entrenched and burdensome, particularly in areas where valuable resources are found. The violence in the concessions is, however, now largely regulated.

In contrast, the Karen living in the rugged hills that cover approximately 75 per cent of Nyaunglebin District have endured near-constant warfare, suffering and loss for the past five decades (KHRG 2001a). Despite the existence of a verbal cease-fire agreement, *tatmadaw* battalions have continued to carry out counter-insurgency operations in remote areas where the Karen National Liberation Army (KNLA, the military wing of the KNU) still exercises some *de facto* control. Since January 2004, when the informal cease-fire began, at least seven more villages have been abandoned out of fear or destroyed by the *Tatmadaw*. *Tatmadaw* units have also built seven new military bases in the district using forced labour.

In 2005, renewed military operations in Kyauk Kyi Township, which borders Shwegyin to the north, displaced another 5,900 civilians, raising the total number of internally displaced people in eastern Pegu (Bago) Division to 21,300 (TBBC 2005:23–4). An unknown number of people died during this period—directly from the conflict, through mistreatment by soldiers, or from hunger, injuries and treatable diseases. The most recent offensive, which began in March 2006 in the northern and western parts of the district, will undoubtedly raise the number of displaced people by as much as another 22,000 (TBBC 2006:22, 37–9). Surprisingly little is known, however, about the forms of governance in the transition zone between the plains and the hills, where extractive industries have long operated. The next section begins to address this gap by focusing attention on the district's waterways, which connect the plains and the hills.

Waterways as governable spaces

In addition to the Shwegyin River, eight rivers and creeks are found in Nyaunglebin District, the largest being the Baw Ka Hta, the Bo Lo and the Mawtama. The importance of these navigable waterways is several-fold. First, the networks of waterways serve as the primary means by which people, food, information and basic commodities circulate within the region. Areas where two rivers meet are particularly important since they form spaces where it is cost effective for armed groups to extract rents and to assert their authority (Scott 1998).[4] For this reason, the *Tatmadaw* and the KNU/KNLA have fought one another for decades to control these important transportation routes and to tax those who use them.[5]

Second, nearly 250 commercially valuable orchards and gardens are located along the Shwegyin and Pa Ta Law Rivers alone.[6] The rich alluvial soils that line these rivers are the basis of the district's food supply and economy. Indeed, many of the cash crops cultivated in Nyaunglebin District are important for regional markets and supply chains that extend to Rangoon (Yangon) and other urban areas. Key crops include betel-nut, durian, rubber and, especially, *shauq-thi*, a type of lemon-lime used widely in Burmese cuisine. The district produces nearly one-third of the country's *shauq-thi* supply, which currently sells for 15 kyats per fruit in Shwegyin and as much as 50 kyats a fruit in Rangoon.[7] These cash crops, however, are currently threatened by a multi-purpose hydroelectricity project being constructed on the Shwegyin River. The dam's reservoir is expected to submerge a significant number of these orchards, which will have a devastating impact on the region's economy.

Third, nearly all of the gold deposits in the district are located in its alluvial soils. Since the late nineteenth century, local inhabitants have extracted gold from these areas using a range of low-impact techniques, primarily to augment their incomes during the dry season. In contrast, large-scale gold-mining operations, which used Chinese-manufactured hydraulic equipment, appeared only in 1997 during the third wave of forced relocations, yet the equipment has devastated much of the area within the space of only a few years. The units use diesel generators to pump water through hoses at extremely high pressure, which is then directed at the banks of rivers and streams to dislodge soil, rock and other sediments to expose ore-bearing layers. The resulting slurry, after passing through variable-sized screens to sort the particulates, is commonly treated with mercury to amalgamate the gold. These techniques, while efficient, are extremely destructive and have entirely drained some water sources and permanently altered others. Many of the district's remaining rivers and streams are now so polluted from the chemical run-off, including acid mine drainage, that Karen farmers report that their fruit and citrus trees are beginning to die. The pollution has also contributed to outbreaks of malaria, dysentery and infectious skin diseases among miners and *tatmadaw* troops stationed at the mining sites, as well as people living further downstream (ERI and KESAN 2003:57–8).

Fourth, as noted above, the Shwegyin River is the site for a multi-purpose, 75-megawatt hydroelectric dam, one of 11 being built in Pegu (Bago) Division (ASEAN India 2005). The Ministry of Electrical Power Enterprise announced in late 2000 that it would oversee the construction of the dam next to the village of Kyaut Nagar, several kilometres outside the town of Shwegyin. The Myanmar Electric Power Enterprise, a state-owned utility responsible for the generation, transmission and distribution of electricity throughout the country, is expected to manage the dam on its completion. Although construction began in 2001, the project has been delayed repeatedly due to heavy seasonal rains and corruption, especially the theft and adulteration of cement. The corruption eventually became so severe that in 2004 the SPDC replaced the state-owned companies contracted by the Department of Electrical Power in Pegu (Bago) Division with two privately owned ones: Olympic Company Limited and Min A Naw Ya Ta.[8] The dam was scheduled for completion in late 2006, but further delays were expected since work on the intake pipe, spillway and power station had yet to begin.[9] Despite these problems, the Shwegyin Township Peace and Development Council announced plans to transform the low-lying plains east of the Sittaung River into vast rubber plantations beginning in late 2006 using water diverted from the dam's reservoir. According to a report in *The New Light of Myanmar* (Anonymous 2005), the plantations will ultimately cover between 50,000 and 100,000 acres by 2009. An unknown number of small landholders who cultivate crops in this area will be displaced as a result.

Finally, gold-mining companies have since 2004 sought to obtain the *Tatmadaw*'s permission to move into the headwaters of the Shwegyin River and its tributaries once the dam becomes operational. If this occurs, their migration will spread the environmental devastation connected to the mining upstream into the Ku Shaw region where Nyaunglebin District and Papun District (Karen State) meet. This region forms the western edge of the Kayah-Karen/Tenasserim Moist Forests, which is listed as one of the world's 200 most significant eco-regions in terms of its biodiversity (WWF 2002). The KNU has repeatedly threatened to attack the mining operations should they move towards Ku Shaw, a response that could cause the already tenuous cease-fire between it and the SPDC to completely collapse.[10] Despite these threats, Military Operations Command 21 established a new headquarters in Ku Shaw in mid 2006 to carry out counter-insurgency operations in the hills between it and the Mawtama River, which suggests the expansion of gold-mining into this area is likely in the near future.[11]

The above developments share two overlapping features that reinforce one another over time to create the means through which authority is constituted and rule accomplished in Nyaunglebin District. The first is the militarisation of everyday life. Since 1999, when the third wave of forced relocations ended, the *Tatmadaw* has constructed (using forced labour) 17 army camps and 25 SPDC-controlled 'relocation centres'. Most are found along the road linking Shwegyin, Kyauk Kyi and Tantabin or along the banks of the Shwegyin River where the concessions and dam are located (TBBC 2005). The second entails the regulated forms of violence that govern how different *tatmadaw* battalions, mining companies, private businessmen and construction firms compete and collude with one another to obtain land and rents from those who reside in the area. Caught in the middle are the now predominantly Burman migrant labourers, who extract the gold, and the Karen, whose economic livelihoods and way of life, based largely on horticulture and petty trading, are being destroyed. 'What once was considered our treasure has now become our sorrow,' said one Karen farmer.[12] Another displaced farmer echoed these sentiments: 'When

the next generation is asked where their parents lived, they will not
be able to say anything because the land will have been destroyed and
there won't be anything left to show them.'[13]

Together, these trends, which are accompanied by severe travel
restrictions, fear and worsening pollution, mean that options continue to
narrow for most Karen living along Nyaunglebin District's waterways. The
only alternatives are to seek employment as day labourers for the extractive
industries operating in the district, to become internally displaced people
or to flee the region entirely. None of these options are attractive and
each carries its own risks, including further abuses, increased morbidity
and premature death (TBBC 2004; 2005). Since the abuses connected to
militarisation are well documented, the sections that follow focus instead
on the regulated forms of violence in Shwegyin Township.

Regulated violence

Most attempts to understand the reasons for displacement in eastern
Burma have for entirely justifiable reasons focused on state-sponsored
violence. Between December 1996 and 2005, the *Tatmadaw* and its
armed proxies forcibly displaced approximately one million people in
this part of the country alone. More than 2,500 villages were destroyed,
relocated or abandoned as a result. Despite cease-fire agreements with
nearly all of the armed groups operating along the country's eastern
border, there are still more than half a million internally displaced people
in this region (TBBC 2005). The scale and severity of this violence,
however, has directed attention away from other forms of displacement
that do not neatly fall into categories of those caused by armed conflict
or large-scale development projects, such as the construction of rail lines
or natural gas pipelines (see Robinson 2003).

This is not to suggest that violence is absent in and around the
concessions. The research team collected troubling accounts of physical
and sexual assault, torture, murder and illegal forms of military
conscription. But the number and severity of these incidents appears
to have decreased in areas where natural resource extraction takes place,

a trend that has been confirmed elsewhere in eastern Burma (TBBC 2005:26, 43). Similarly, the *Tatmadaw's* use of forced labour, although still widespread across the country (ILO 2006), has largely disappeared in the mining concessions as well. Thus the concessions appear to regulate violence by establishing semi-formal, if still largely arbitrary and extra-legal, rules, which help govern access to particular resources and the kinds of rents that can be levied by different actors. The two primary means for doing so in Shwegyin Township are outlined in the sections that follow.

Tactics for acquiring property

The extractive industries and *tatmadaw* battalions operating in Shwegyin Township have utilised a number of different strategies to secure access to land and steady streams of revenue. Since different resources are often found in the same spaces, conflicts over how extraction is to occur and under whose control are not uncommon. The main strategies, which have changed over time in response to this competition, can be subdivided into two categories: those related to the extractive industries and those related to different kinds of infrastructure projects in the township, the largest of which is the Kyaut Nagar Dam. Although significant differences exist between strategies used, they both result in similar outcomes: displacement.

According to local sources, efforts to intensify gold-mining in Shwegyin Township began as early as 1995; however, mining companies did not begin to seek to secure land rights on a large scale until 1997. The initial expansion coincided with the wave of forced relocations the Light Infantry Division No.77 was then carrying out along the district's waterways. During the first years of the 'gold-rush', in which an estimated 10,000 people arrived in the district, many enterprising miners simply offered large sums of money to whomever was there cultivating the land and caring for the orchards at the time (ERI and KESAN 2003:17). This strategy often worked as previous waves of violence had led many of the *de jure* landowners to relocate elsewhere.

In many instances, it was impossible to know whether the real owner was still alive and, if so, where he or she might be located—given the frequency with which Karen living in the area have had to move to avoid state-sponsored violence or in direct response to it (Burma Border Consortium 2003:47, 50). When people did return, many found their land occupied by kin, friends, neighbours or other internally displaced people, which was a source of considerable inter-personal conflict.[14]

For those who remained, the unceasing demands from *tatmadaw* units for forced labour, taxes, food and other materials ensured that most Karen in the area remained poor and hungry (Burma Border Consortium 2003; AHRC 1999). Given these conditions, it is not surprising that some landowners opted to sell property, which may or may not have been their own, to representatives of the mining companies when offered to purchase them. While the sums were still many times below real market value, they were nonetheless considerable. Reported amounts ranged anywhere from 500,000 up to three million kyats.[15] Even then, this was many times below market value. Gold-mining companies shifted strategies in 2001 after construction on the dam began. Since the dam's reservoir would eventually flood the area, the mining companies opted to work directly with *tatmadaw* battalions to convince Karen landowners to sell their property at below market prices or face violent retribution. (The shift in tactics also led to the gradual consolidation of the industry, which by 2003 was dominated by three companies, each of which employed between 1,000 and 1,500 labourers: Aye Mya Pyi Sone, Kan Wa and Ka Lone Kyeik.[16] Since most Karen, like other non-Burmans, lack national identification cards and/or a full set of title deeds to their property (TBBC 2005), it proved easy to use these quasi-legal, albeit unjust and highly discriminatory, tactics to force a sale. To date, nearly all of the original inhabitants of Ywa Myoe, Kun Nie, Be La, Htee Ka Hta, Ta Nee Pa and Su Mu Hta villages—a predominantly Karen area—have sold or abandoned their fields due to these tactics.[17] Nonetheless, the appearance of legality remains important. As one indication, the forcible seizure of land without compensation remains rare, even in areas where significant gold deposits are located.

In contrast, *tatmadaw* units based at the dam site and the construction companies have expropriated large areas of land around Kyaut Nagar since 2001. In most instances, homes were destroyed with bulldozers with little or no advance warning. No compensation has been provided for those displaced by the project.[18] Additionally, to obtain sufficient fill for the dam, which is 1.1km long and 56m high, Min A Naw Ya Ta and the Olympic Company Limited are using heavy equipment and large teams of day labourers to remove rocks from the bed of the Ka Tee Chong River. The rocks are then transported to the dam site by truck. Some enterprising local businessmen have also hired their own teams of Burman day labourers to find stones closer to the construction site, which typically involves trespassing on private property and causing damage to fields and orchards. The use of forced labour by Light Infantry Brigades Nos 20 and 57 is also common at the dam site and in connection with other infrastructure projects, again in contrast with the mining concessions, where wage labour is the norm.[19] Forced labour is particularly acute where roads are being either upgraded or built anew. This expanding road network has permitted logging companies to extend their operations into more remote areas of the district. The roads have also enabled the *Tatmadaw* to sustain its supply lines during the rainy season, allowing it to continue a military offensive year-round for the first time. Forced relocations continue to occur as well.[20] In 2005, a total of six villages were affected; two villages located near the dam site were destroyed, while the residents of another four villages were forced to relocate as punishment for having contact with the KNLA.[21]

The continuing patterns of abuse are contributing to a growing sense of 'resource fatalism' (Inbaraj 2004). According to local sources, everyone feels increasingly compelled to participate in the destruction of their ecosystems in order to earn some income before everything of economic value is extracted.[22] A Karen man, who lost his orchard to mining companies and now makes charcoal to help pay the fees imposed on him by the *Tatmadaw*, explained: 'If we do not burn charcoal, we will not be able to eat. But if we do burn the charcoal, it will affect the environment. When all the trees are gone, we do not know what we will do' (KESAN 2003:25). Since Karen residents lack recourse to legal remedies, the only available alternative is to stand aside and watch others consume the basis of their economic livelihoods.

Tactics for acquiring rents

The militarisation that has accompanied the expansion of gold-mining ventures is not simply a means to provide security for these operations. The presence of large numbers of soldiers has also permitted the military to strengthen control over the local economy by extracting an array of rents, which are commonly defined as the extraction of uncompensated value from others. The predatory practices employed by *tatmadaw* battalions and their troops offer a case in point. Before the 2006 offensive, there were five battalions based in Shwegyin Township: Light Infantry Brigades Nos 589, 598, 349, 350 and Infantry Brigade No.57. Since each of these battalions has to extract rents to help cover operating costs and to make regular payments to their superior officers, the field units compete with one another while stationed temporarily in the concessions to extort additional resources from those who live or work in the extractive zones.[23] (A similar system of 'gates'—that is, military check-points where additional rents are extracted—is in place along the district's roads and rivers but is not discussed here.) This competition also extends down to those units drawn from the same 'mother' battalion. Miners and farmers, for example, report that units rotate every month and demands for non-scheduled rents typically occur just before the soldiers return to their base, a practice that decreases the amount of money and food available to those who arrive to take their place. According to local sources, most of the illicitly gained income moves up the chain of command, first to Brigadier-General Thura Maung Nyi, who heads Division No.77, and then to Major-General Ko Ko, who is in charge of the Southern Regional Command based in Toungoo.[24]

The rents, although they generate fairly predictable revenue streams, are modest compared with the income generated by the gold itself. During

Table 11.1   Selected list of rents extracted in the mining concessions, Shwegyin Township, 2004–2005

Collected by	Type	Amount (kyat)
Tatmadaw units	Security fee for mining companies	10–20,000 per month (plus additional fees)
Tatmadaw units	Tax for small-business operators (tea, video, karaoke and casino/brothel) operating within concessions	400 a night
Tatmadaw units	Security fee for small-business operators	1,500–3,000 per shop per month
Tatmadaw units	Residence tax for miners and dependents	700 per person per month
Tatmadaw units	Travel fee to enter and exit concessions	500 per person (valid one week to one month)
Tatmadaw units	Security fee for landowners near mining sites	1–2,000 per owner per month
Tatmadaw units	Permit fees for firewood collection	3,000 per person per month
Mining company	Scavenging fee	2–3,000 per person per day to search tailings for gold

High-ranking military officials and businessmen	Lease fee for mining on private property	Landowner retains 60 per cent of all gold extracted
Tatmadaw battalions	Tax on miners employed by company	1,000 per miner per month
Division headquarters	Concession fee (separate from amount paid to Department of Mines)	100,000 per month
Division headquarters	Rental fee for hydraulic equipment (goes to 'Division Fund')	1–500,000 per maching per month (varies according to productivity of site)
Brigadier-General Thura Maung Nyi	Permission fee paid by battalions to collect the above	500,000 per month

Source: Field Survey No.001 (2002); Interview Nos 001–5 (2005), 001–2FU (2005)

the early years of the gold-rush, high-ranking *tatmadaw* commanders, such as General Thura Maung Nyi, and local businessmen, such as Po Baing, reportedly purchased large tracts of land. Under current leasing arrangements, these men retain 60 per cent of all the gold extracted on properties they own. According to local sources, each of the mining concessions (which frequently has more than one hydraulic unit operating within it) produces an estimated 1.5kg of gold dust and flakes a month.[25] While not a significant amount by international standards, the price of gold in Burma has nonetheless increased dramatically in the past several years, making it a highly lucrative source of income. In 2003, one *kyat thar* of gold (1.53 grams) sold for 90,000 kyats in Shwegyin Township (ERI and KESAN 2003:55).[26] By April 2006, concerns about the stability of the regime and related inflationary pressures drove the price to a record high of 500,000 kyats per tical (16.3g), according to *Xinhua* (21 April 2006). At these rates, 1.5kg of gold a month would provide an astronomical return, especially when compared with the gross domestic product per capita in Burma, which is estimated to be a mere US$1,700 (CIA 2004).

These findings, which are summarised in Table 11.1, support a working hypothesis that the fee structure is not simply an extra-legal means to create wealth.[27] The system, which fosters competition within and between different segments of the *Tatmadaw*, appears to establish a framework where ambitious officers, by strategically redistributing goods and services (including rights to collect rents), can advance their careers (see also KHRG 2001a). The arrangements also suggest that regulatory controls on the extractive industries are, at best, weak (see Gutter 2001). At worst, the staff members who work for the government departments representing different ministries are complicit in the abuses occurring around them. The Department of Mines, to offer one example, is charged with implementing the terms of the 1994 Myanmar Mines Law. Among other things, the law requires permit-holders to create safe conditions for workers; to use land and water in accordance with existing laws; and to pay royalties of between 4 and 5 per cent on all gold extracted (SLORC 1994). These and other requirements specified under the law, however, appear never to have been enforced by the Department of Mines. In fact, local informants make no reference to the department in any of their accounts.

Conclusion

What accounts for or contributes to these forms of government along the waterways of Nyaunglebin District? In part, living conditions throughout much of the country have deteriorated to the point that, in order to feed themselves, many people now have to participate in practices that are morally corrosive and result in environmental degradation. One clear sign of this can be found in the concessions themselves. While some of the labourers working along the riverine networks are from the area, the majority of the miners are landless, economic migrants (ethnic Burmans, Shan and Chinese), who opt to perform the work despite the dangers involved and the fact that pay scales are rarely sufficient to meet daily expenses given high rates of local inflation. This emergent proletarian class, however, remains internally subdivided as all of the extractive industries exhibit a strong preference for hiring workers from their own ethnicity and men earn considerably more than women, even when they perform the same task. Additionally, Karen are employed only as a last resort, which leaves commercial charcoal production and the harvest of rattan and bamboo as the few ways subsistence farmers and internally displaced people can earn cash to purchase medicine, cooking oil and other necessities (ERI 2003). This is not to imply that people who live and work in the concessions lack agency; rather, it is to convey that for Burmans and non-Burmans alike, their scope for action, especially for anything beyond mere survival, has become sharply circumscribed in the past decade (Heppner 2005; Cusano 2001; TBBC 2005; Agamben 1998).

More generally, this case study has emphasised some of the key processes that make former conflict zones governable, namely: the militarisation of everyday life and the regulated violence that has accompanied it. The findings reveal that contradictory modes of government currently exist along the waterways of Nyaunglebin District, especially in Shwegyin Township, where mining concessions overlap those areas affected by the construction of the Kyaut Nagar Dam. The various actors involved—*tatmadaw* battalions, mining companies, construction companies and related state agencies—have devised quite different strategies and techniques for constituting their authority and for disciplining their labourers, even though they operate along the same short stretch of the Shwegyin River. Some of these actors have found ways to collaborate with each other, whereas other actors continue to work at cross purposes. Still other actors manage to do both simultaneously: the fierce competition for rents not only pits *tatmadaw* battalions against one another, but against their own soldiers, who seek to extract yet more resources from the mining companies and their labourers before rotating out of the mining concessions.

Such practices have simultaneously extended and fragmented the centralised state control of spaces where natural resources are located. This apparent contradiction is made possible by revenue flows that generate and sustain powerful patron–client networks that cross-cut the boundaries imagined to separate state from non-state institutions (see Nordstrom 2000, 2006; Roitman 1998). So while the military, administrative and economic reach of the regime is clearly growing stronger in Nyaunglebin District, the means by which its authority is exercised remains far from coherent or benign. The dam, as will be recalled, will displace the gold-mining companies, providing them with the opportunity to extend their operations towards Ku Shaw—a move that will likely lead to further armed conflict between the *Tatmadaw* and the KNU, as well as environmental degradation. Additionally, the dam will enable a massive commercial plantation east of the town of Shwegyin to be created. If this occurs, several thousand more farmers will lose their land and the district's economy, once based on a diverse array of crops, will be replaced by a single, non-edible commodity: rubber. These cascading forms of displacement, which follow those that have already forced most of the Karen population that once lived along Nyaunglebin District's waterways to leave (MacLean forthcoming), indicate that the transformation of Shwegyin Township is far from over.

Notes

1 For a discussion of the differences between 'governance' and 'governmentality', see Rose 1999:15–24. The emphasis here will be on governance.
2 Other important resources include timber, charcoal, bamboo and rattan.
3 Interview Nos: 003 (2002), 002GM (2004), 001, 003, 002FU (2005). Unless noted, all interviews were conducted by EarthRights International and are on file with the organisation. To protect informants' identities, interviews as well as field documents and surveys are referred to by number rather than name.
4 Scott referred to such sites as potential 'state spaces', however, they are equally open to control by other (armed) groups, as is the case here.
5 The Democratic Karen Buddhist Army (DKBA), a splinter group allied with the SPDC, conducted itself similarly, but as of mid 2005 is no longer officially active in the district.
6 Field Survey No. 002 (2004).
7 Interview Nos 001, 006 (2005).
8 Field Document Nos 003, 006 (2005).
9 Field Document Nos 003, 006 (2005)
10 Opinions remained divided on the broader significance of the 2006 offensive. Some argue that the cease-fire still holds and that the operations were narrowly intended to create a larger buffer around the new capital in Naypyitaw, to punish anti-cease-fire factions within the KNU (especially Brigade No.2), and to secure access to natural resources in areas still patrolled by the KNLA. Others assert that the offensive is really a precursor to a much larger one intended to bring all of Nyaunglebin District and northern Karen State firmly under Rangoon's control. Among other things, this would facilitate the completion of large hydroelectricity projects along the Salween River.
11 Interview Nos 005–6 (2005).
12 Interview No. 006 (2005).
13 Interview No. 006 (2005)
14 Interview Nos 002FU (2005), 003 (2005); Field Document No.009 (2005).
15 Field Document No. 002 (2004).
16 Interview No. 003 (2005).
17 Field Survey No. 001 (2002); Interview No.002FU (2005).
18 Interview Nos 001, 004 (2005).
19 Interview Nos 002, 004 (2005).
20 Field Document No.009 (2005); Interview No.118 (2004); ERI 2005:17–18.

21 Interview Nos 003–5 (2005); Field Survey No. 004 (2005); TBBC 2004:67, 2005:80.
22 See also Interview Nos 001–2GM (2004), 002, 005 (2005); ERI and KESAN 2003:28–35; KESAN 2003:25, 28.
23 Field Survey No. 001 (2002); Interview Nos 001–5 (2005), 001–2FU (2005).
24 Interview No. 117 (2004). DKBA units have reportedly posed as *tatmadaw* troops in the past to collect fees from miners, which results in the same problem. Interview Nos 001–2FU (2005).
25 Local sources typically refer to the number of hydraulic mining machines operating in a general area, such as a stretch of river, instead of concessions *per se*. Between 20 and 30 hydraulic units have been operating along the Shwegyin River between the dam site and Kan Nee since 2003. Other sources, however, place the total closer to 50 machines. These figures do not include machines on other nearby tributaries, for example, the Mawtama River, where mining is similarly widespread, and parts of Kyauk Kyi Township, where at least 40 machines were operating in 2004. Interview Nos 117 (2004), 004, 007 (2005); Free Burma Rangers, email communication with author, 19 April 2006 and 13 May 2006.
26 Field Document No. 002 (2005).
27 Field Survey No. 001 (2002); Interview Nos 001–5 (2005), 001–2FU (2005).

References

Agamben, G., 1998. *Homo Sacer: sovereign power and bare life*, Stanford University Press, Stanford.

Anonymous, 2005. 'Chairman of the State Peace and Development Council, Commander-in-Chief of Defence Services Senior General Than Shwe inspects Shwe Kyin Hydel power plant project', *The New Light of Myanmar*, 12 March, as cited in *Myanmar E-News Letter*, 14 March:3.

——, 2006. 'Myanmar seeks ways to cope with rising commodity prices', *Xinhua News Service*, 30 March 2006, reprinted in *Burma Net News*, 30 March. Available from http://www.burmanet.org/news/2006/03/30/ (accessed 12 July 2006).

ASEAN India, 2005. *Country Profile: Myanmar, opportunities in the electricity sector*. Available from http://www.aseanindia.net/asean/countryprofiles/ myanmar/hydro-electric.htm (accessed 12 July 2005).

Asian Human Rights Commission, 1999. *Voice of the Hungry Nation: the people's tribunal on food scarcity and militarization in Burma*, Asian Human Rights Commission, Hong Kong.

Bahro, R., 1978. *The Alternative in Eastern Europe*, Verso, London and New York.

Burma Border Consortium, 2003. *Reclaiming the Right to Rice: food security and internal displacement in Eastern Burma*, Burma Border Consortium, Bangkok. Available from http://www.ibiblio.org/obl/docs/BBC-Reclaiming_the_Right_to_Rice.pdf

Burma Ethnic Research Group and Friedrich Naumann Foundation, 1998. *Forgotten Victims of a Hidden War: internally displaced Karen in Burma*, April, Burma Ethnic Research Group, Chiang Mai. Available fromhttp://www.ibiblio.org/obl/docs3/Berg-Forgotten_Victims.pdf.

Central Intelligence Agency, 2004. *The World Fact Book: Burma*. Available from https://www.cia.gov/cia/publications/factbook/geos/bm.html (accessed 12 June 2006).

Cusano, C., 2001. 'Burma: displaced Karens: like water on the *khu* leaf', in M. Vincent and B.R. Sorensen (eds), *Caught Between Borders: response strategies of the internally displaced*, Pluto Press, London:138–71.

EarthRights International and Karen Environmental and Social Action Network, 2003. *Capitalizing on Conflict: logging and mining in Burma's cease-fire zones*, EarthRights International, Chiang Mai.

EarthRights International, 2005. '*If We Don't Have Time to Take Care of Our Fields the Rice Will Die': a report on forced labor in Burma*, EarthRights International, Chiang Mai. Available from http://www.earthrights. org/files/Reports/ILO_ForcedLaborReportinBurma2005.pdf.

——, 2007. *Turning Treasures into Tears: mining, dams and deforestation in Shwegyin Township, Pegu Division*, EarthRights International, Chiang Mai. Available from http://www.earthrights.org/files/ Burma%20Project/report-_turning_treasure_into_tears.pdf.

Ferguson, J., 2006. 'Governing extraction: new spatializations of order and disorder in neoliberal Africa', in *Global Shadows: Africa in the neoliberal world order*, Duke University Press, Durham:194–210.

Foucault, M., 1991. 'Governmentality', in G. Burchell, C. Gordon and P. Miller (eds), *The Foucault Effect: studies in governmentality*, Harvester Wheatsheaf, London: 87–104.

Frynas, J.G., 1998. 'Political instability and business: focus on Shell in Nigeria', *Third World Quarterly*, 19(3):457–78.

Global Witness, 2003. *A Conflict of Interests: the uncertain future of Burma's forests*, Global Witness, London.

Gutter, P., 2001. 'Environment and law in Burma', *Legal Issues on Burma* 9:1–25.

Hardin, R., 2002. 'Concessionary Politics in the Western Congo Basin: history and culture in forest use', Environmental Governance in Africa Working Papers Series No. 6, November, World Resources Institute, Washington, DC. Available from: http://pdf.wri.org/eaa_wp6.pdf

Heppner, K., 2005. *Sovereignty, Survival, and Resistance: contending perspectives on Karen internal displacement in Burma*, Karen Human Rights Group, Kawthhoolei. Available from http://www.khrg.org/papers/wp2005w1.pdf (accessed 2 January 2006).

Houtman, G., 1999. *Mental Culture in Burmese Crisis Politics: Aung San Suu Kyi and the National League for Democracy*, Institute for the Study of Languages and Cultures of Asia and Africa, Tokyo University of Foreign Studies, Tokyo.

Inbaraj, S., 2004. 'China covets Burmese resources', *The Irrawaddy*, 27 July.

Images Asia and Pan Kachin Development Society, 2004. *At What Price: gold mining in Kachin State, Burma*, November, Nopburee Press, Chiang Mai. Available from http://www.ibiblio.org/obl/docs/gold%20pdf1.pdf.

International Crisis Group, 2001. *Myanmar: the military regime's view of the world*, Asia Report No. 28, December, International Crisis Group, Bangkok.

International Labour Organization, 2006. *Conclusions on Document GB.295/7: developments concerning the question of the observance by the government Myanmar of the forced labour convention, 1930 (No.29)*, International Labour Organization, Geneva.

Karen Environmental and Social Action Network, 2003. *Thulei Kawwei* [Karen Environmental Forum], Karen Environmental and Social Action Network, Chiang Mai.

Karen Human Rights Group, 2001a. *Flight, Hunger, and Survival: repression and displacement in the villages of Papun and Nyaunglebin Districts*, Karen Human Rights Group, Kawthoolei. Available from http://www.khrg. org/khrg2001/khrg0103a.html.

Karen Human Rights Group, 2001b. *Abuse Under Orders: the SPDC and DKBA soldiers through the eyes of their soldiers*, Karen Human Rights Group, Kawthoolei. Available from http://www.ibiblio.org/freeburma/ humanrights/khrg/archive/khrg2001/khrg0101.pdf.

——, 2005. *Proliferation of SPDC army camps in Nyaunglebin District leads to torture, killings and landmine casualties*, 7 July 2005. Available from http://www.khrg.org/khrg2005/khrg05b5.html (accessed 22 August 2006).

Karen Rivers Watch, 2004. *Damming at Gunpoint*, Karen Rivers Watch, Kawthoolei.

MacLean, K. and Mahn Nay Myo, 2002. 'Forced labor on the Shwegyin River in Burma', *World Rivers Review*, 17(4):11.

MacLean, K., (forthcoming). 'Communities in forced motion: concessions, state-effects, and regulated violence in Burma', in N. Peluso and J. Nevins (eds), *Taking Southeast Asia to Market: commodifications in a neoliberal age*, Cornell University Press, Ithaca.

Nordstrom, C., 2000. 'Shadows and sovereigns', *Theory, Culture, and Society*, 17(4):35–54.

——, 2006. *Global Outlaws: crime, money, and power in the contemporary world*, University of California Press, Berkeley.

Ohmae, K., 1995. *The End of the Nation-State: the rise of regional economies*, Harper Collins, London.

Robinson, C.W., 2003. *Risks and Rights: the causes, consequences, and challenges of development-induced displacement*, Brookings Institute and CUNY Project on Internal Displacement, New York.

Roitman, J., 1998. 'The Garrison-entrepôt', *Cahiers d'Etudes africaines*, 38(150–2):297–329.

Rose, N. 1999. *Powers of Freedom*, Cambridge University Press, Cambridge.

Scott, J., 1998. 'Freedom and freehold: space, people, and state simplification in Southeast Asia', in D. Kelly and A. Reid (eds), *Asian Freedoms: the idea of freedom in East and Southeast Asia*, Cambridge University Press, Cambridge:37–64.

State Law and Order Restoration Council, 1994. *The Myanmar Mines Law*, 6 September 1994. Available from http://www.ibiblio.org/obl/docs3/Mining%20Law.htm (accessed 12 July 2006).

Thai Burma Border Consortium, 2004. *Internal Displacement and Vulnerability in Eastern Burma*, October, Thai Burma Border Consortium, Bangkok. Available from http://www.ibiblio.org/obl/docs/TBBC-IDPs2004-full.

——, 2005. *Internal Displacement and Protection in Eastern Burma*, October, Thai Burma Border Consortium, Bangkok. Available from http://www.ibiblio.org/obl/docs3/TBBC-Internal_Displacement_and_Protection_in_Eastern_Burma-2005.pdf.

——, 2006a. *Internal Displacement in Eastern Burma: 2006 survey*, Thai Burma Border Consortium, Bangkok. Available from http://www.ibiblio.org/obl/docs4/TBBC-2006-IDP-ocr.pdf.

——, 2006b. *Internal Displacement in Eastern Burma: July 2006 update*, Thai Burma Border Consortium. Available from http://www.tbbc.org/news/idp-update-06-07-28-page.html (accessed 22 August 2006).

Tse, P.K., 1991. 'The mineral industry of Burma', *Minerals Yearbook: mineral industries of Asia and the Pacific*, (3):60–9, US Bureau of Mines, Washington, DC.

Watts, M., 2004. 'Resource curse? Governmentality, oil, and power in the Niger Delta, Nigeria', *Geopolitics*, 9(1):50–80.

World Wildlife Fund, 2002. *The Global 200*. Available from http://www.nationalgeographic.com/wildworld/profiles/g200_index.html (accessed 14 July 2006).

12 Identifying conservation issues in Kachin State

Tint Lwin Thaung

Kachin State in northern Myanmar is home to many biological hotspots, including subtropical moist forests, hill forests, alpine meadows and broadleaf and conifer forests (Olson and Dinerstein 1998). Global Witness (2005) recently reported considerable unease about the scale of illegal forest activities in Kachin State. Kahrl et al. (2004) analysed the China–Myanmar timber trade and its implications for forests and livelihoods in Myanmar's Kachin State and the Yunnan Province of China. They found that China's demand for timber was an underlying cause for the unsustainable harvest of valuable forests in Kachin State. Unsustainable logging was discussed comprehensively in the above-mentioned studies, but the views of local stakeholders from Kachin State were not thoroughly considered. This chapter seeks to understand the views of local stakeholders in regard to natural resource conservation issues.

This chapter discusses data resulting from a study complementary to an earlier one by Webb et al. (2004). Findings and analysis in this previous study were based on a literature review, remote-sensing data and stakeholder interviews in Yangon and Mandalay. They revealed a wide scope of conservation issues reflected at the national level. This present chapter seeks to verify the results of the 2004 study with reference to real situations occurring on the 'front line' or at local levels.

Methodology

Participatory rural appraisal (PRA), one of the methodologies of farmer participatory research, was used to generate data for the present study. This study employed semi-structured interviews with local stakeholders and direct observation. Working closely with local stakeholders helped to determine local conditions, perceptions and preferences in conserving natural resources in Kachin State. A semi-structured interview schedule was prepared with the list of 'incompatibilities' used by Rao et al. (2002) in evaluating the protected area system in Myanmar. These incompatibilities were renamed as threats or issues in this study, and stakeholders were welcome to freely raise other, unlisted issues.

The stakeholders who participated in the previous SWISSAID Myanmar program participated actively in the interviews. Local stakeholders who lived or worked in Kachin State were categorised as academics, non-governmental organisation (NGO) workers, businesspeople and those from peace groups or State Peace and Development Council (SPDC)-designated 'national races groups' and

Table 12.1   Categories of stakeholders interviewed

NGOs	Local businesspeople	Academics	Ethnic armed groups
SWISSAID	Jade-mining	Institute of Forestry	Kachin Independence Organisation (KIO)
World Concern	Photography	Zoology	
YMCA	Traditional medicine	Botany	
Shalom			

Source: Author's compilation

government agencies. Thirteen stakeholders from those categories were requested to identify important conservation issues (Table 12.1).

Conservation issues were separated into large-scale (Table 12.2) and small-scale (Table 12.3) issues. The issues were then ranked by the stakeholders in a way that was reflective of their individual perceptions of constraints to conservation in Kachin State. Impacts of threats and the prevalence of such threats were also ranked.

Table 12.2 Large-scale issues arising from official projects with institutional support and driven by larger commercial interests

Stake holder	Hunting			Firewood			NTFP			Grazing			Fishing			Shifting cultivation		
	T	I	P	T	I	P	T	I	P	T	I	P	T	I	P	T	I	P
1	2	3	2	3	3	3	3	2	3				2	3	3	3	3	3
2	2	2	2	3	3	3	2	2	2	1	1	1	2	2	2	2	3	2
3	2	2	3	3	2	3	3	1	3	2		2	3	3	3	3	3	3
4	3	3	2	3	2	3	3	3	3				2	3	2	2	3	3
5	2	3	2	1	1	2	1	1	2	1	1	2				3	3	3
6	3	3	3	3	3	3	1	2	3				3	3	3	3	3	3
7	3	2	3	3	1	3	1	3	3				1	1	2	2	3	3
8	2	2	3	1	3	3	3	3	3				1	1	3	3	3	3
9	3	3	3	2	3	2	2	2	2				2	2	2	3	3	3
10	3	3	3	2	2	2	2	2	2				1	1	1	3	3	3
11	3	3	3	3	3	3	3	3	3				3	3	3	3	3	3
12	2	2	2	1	1	1	2	2	2				2	2	2	3	3	3
13	3	3	3	2	2	2	2	2	2				2	2	2	2	2	2
Rank sum	33	34	34	30	29	33	28	28	33	4	2	5	24	26	28	35	38	37

Notes: T threats (3: the most serious; 2: very serious; 1: serious)
I impact (3: high impact; 2: moderate impact; 1: low impact)
P prevalence (3: mostly occurred; 2: sometimes occurred; 1: never occurred)
Source: Author's calculations

Table 12.3   Small-scale issues, often driven by financial hardship

Stake holder	Mining			Human settlement			Infrast-ructure			Plantation			Armed conflict			Cultivation			Tourism			Breeding centre			Logging		
	T	I	P	T	I	P	T	I	P	T	I	P	T	I	P	T	I	P	T	I	P	T	I	P	T	I	P
1	3	3	3				2	2	3	1	1	1	3	3	3				1	1	1	1	1	2	3	3	3
2	2	2	2	3	3	3	2	2	2	2	2	2	2	2	2	1	1	1	1	1	1	1	1	1	3	3	3
3	3	3	3	2	3	2	2	2	3				2		3				2	2	2	2		2	3	3	3
4	3	3	3	1		2	2	4	2	2	1	2	3	3	3										3	3	3
5	3	3	3	1	2	2	3	2	2	3	3	3	3	3	3	1	1	2							2	3	3
6	3	3	3	2	1	2	1	3	3				2	3	3	2	2	3	1	1	2				3	3	3
7	2	3	3	2	2	3	2	3	3				1	2	2	2	1	2	1	1	2				2	3	3
8	1	3	3	2	2	3	1	1	2	2	2	2	3	1	3	1	2	3	2	2	3				3	3	3
9	3	3	3				1	2	1				3	3	3										3	3	3
10	3	3	3	1	1	1	2	2	2				3	3	3	1	1	1							3	3	3
11	3	3	3	1	1	1	3	3	3	1	1	1	3	3	3	1	1	1	1	1	1				3	3	3
12	3	3	3	1	1	1	1	1	1				3	3	3	1	1	1							3	3	3
13	3	3	3	2	2	2	2	2	2				3	3	3	1	1	1							3	3	3
Rank sum	35	38	38	12	12	15	22	23	29	11	9	10	34	32	37	9	9	11	7	7	7	12	4	2	37	39	39

Notes: T threats (3: the most serious; 2: very serious; 1: serious)
I impact (3: high impact; 2: moderate impact; 1: low impact)
P prevalence (3: mostly occurred; 2: sometimes occurred; 1: never occurred)
Source: Authors calculations

Local conservation concerns

The most serious large-scale conservation issues as ranked by stakeholders were logging, mining, the presence of armed groups and infrastructure development (such as road construction). These issues are believed to have a major impact on conservation in Kachin State. Permanent human settlement, industrial plantations and permanent cultivation were ranked as very serious issues with moderate impact, and as occurring occasionally in Kachin State. Tourism was ranked as a low-impact threat, although there was great potential for the development of a tourist industry in Kachin State because of its natural beauty. The stakeholders ranked shifting cultivation, hunting and wood collection as the most serious and widely distributed small-scale conservation issues in Kachin State (Table 12.4). Many small-scale activities (for example, gold-mining) can cause impacts on a scale similar to those of large-scale activities. Fishing and the collection of non-timber forest products are also very serious, though their impact is still low. Grazing is not a common conservation concern in Kachin State.

Some stakeholders raised other relevant factors in efforts to protect the rapid depletion of natural resources in Kachin State. One was the low morale or disempowerment of much of the population, stemming from local and broader issues including corruption and abuse of power. Many people exploit natural resources for reasons of financial survival and are not concerned primarily about long-term livelihoods or sustainable development.

A second issue concerns the complicated governance situation among the different armed groups. Currently, there are four main groups controlling resource exploitation in Kachin State: the government (SPDC or northern military command); Kachin Independence Organisation (KIO), which manages Special Region 2; New Democratic Army—Kachin (NDAK), which manages Special Region 1; and another group recently split from the KIO. In addition, the Pa-O Peace Group plays a major role in resource exploitation, particularly in jade-mining. This complicated and overlapping governance system inevitably causes conflicts over resource exploitation.

Table 12.4   Threats ranked by stakeholders, regardless of scale

Issues	Sum
Timber extraction	37
Shifting cultivation	35
Mining	35
Military/ethnic armed groups	34
Hunting	33
Firewood collection	30
NTFP	28
Fishing	24
Infrastructure development	23
Permanent human settlement	15
Permanent cultivation	9
Tourism	6
Industrial plantations	5
Grazing	4
Breeding centres	4

Notes: 30 or >30 = the most serious threats; 21–9 = very serious threats; 0–19 = serious threats.
Source: Author's calculations

High unemployment rates and associated social welfare issues are expressed as a third important issue relevant to conservation. There are many underlying causes for high unemployment among the local people. The more control is held by influential people with large-scale business activities, the fewer opportunities there are for local people with small-scale ones. For example, jade was a common resource exploited by local Kachin people until jade-mining was recently monopolised by 'peace groups' from other regions. Chinese contractors are using their own labourers even for low-paid jobs in road construction, which they justify by stating that local people are not skilled. Authorities who deal with Chinese contractors have little bargaining power for local employment.

A fourth issue pointed to by stakeholders was China's high market demand. Economic development in China relies to a substantial degree on imported resources. Fuelled by the political influence of the Chinese government and an attractive short-term market, natural resources, especially forest resources in Kachin State, are rapidly disappearing. It has been estimated that because of excessive timber demands from China, the natural forests of Myanmar will be gone in 10–15 years if the current cutting rate continues (Ktsigris et al. 2005).

A final issue raised by stakeholders was the expansion of opium plantations across Kachin State in recent years (Khun Sam 2006). According to a 2005 opium survey by the United Nations Office on Drugs and Crime (UNODC), opium-poppy cultivation in Kachin State had increased in recent years while it decreased in other regions of Burma. Despite recent eradication measures by Burmese authorities, production increased in Kachin State by 900 per cent in 2005, according to the UNODC. The evidence shows that raw opium and other drugs are carried to China concealed among logs.

In our first study (Webb et al. 2004), various stakeholders in Yangon and Mandalay were interviewed using semi-structured forms and open discussion. We listed threats, opportunities and suggestions for research based on the data obtained from interviews, a review of the literature and remote-sensing data (Webb et al. 2004). Some issues investigated in this earlier study proved not to be relevant to Kachin State, and there were significant differences between the tested issues. The serious conservation issues in Kachin State—such as the impact of logging, mining, infrastructure development, shifting cultivation, the presence of armed groups, hunting and wood collection—are, however, common to both studies.

Logging in Kachin State

Unsurprisingly, logging is the most common issue raised by all stakeholders and supported by many reports. There were no opportunities to observe the magnitude of logging in Kachin State while we were there since the newly posted northern military commander had temporarily banned logging. Recently, corrupt officials have been charged and penalised. This has not stopped influential businesspeople and peace group leaders approaching the authorities in order to gain permission to resume logging—permission that has recently been granted.

The underlying causes of logging are complicated, since many powerful stakeholders are involved in illegal practices. Powerful stakeholders include ethnic armed groups, regional military leaders, Chinese business tycoons, drug smugglers and corrupt officials from China and Myanmar. Minority group leaders and the SPDC's northern military command grant logging contracts to Chinese companies as turnkey projects. In return, Chinese companies are the ones building the roads, bridges, power stations, schools and clinics in Kachin State. The only stakeholders who have no voice are local people. They have not seen any tangible benefits from the turnkey projects, and project outcomes are sporadic and fragmentary. In addition, local people have the most to lose from unsustainable or illegal logging practices. Recent flooding in Myitkyina, the capital of Kachin State, is an example of the consequences of unregulated logging and consequent deforestation, and it has devastated the livelihoods of local people. Weak policy, institutions, legislation and infrastructure contribute to illegal logging practices.

In this study group's last trip to the China–Myanmar border in July 2006, the transport of illegal logs across the border at Laiza was observed. At that time, illegal logging was continuing, but less intensively than previously. For example, about 30 trucks of timber were still crossing the border at Laiza daily, compared with more than 100 timber trucks before. This reduction could have been due to unfavourable weather conditions as well as a temporary response to recent crack-downs by Yunnan officials on illegal timber transportation. On Myanmar's side, a northern military commander has been credited for his effort to control illegal logging activities in Kachin State and to provide more freedom in trade by removing many unnecessary check-points. It is difficult to know, however, how long this situation can be maintained, as there are many internal pressures within the military and from their business élites. No conservation issue is harder to solve than the logging of Kachin State's dwindling forests.

Secondary conservation issues

Shifting cultivation has been practised in Kachin State for many years as a form of traditional farming. That shifting cultivation causes deforestation is not new, but due to a growing population and scarce land resources, the practice has passed beyond its traditional scale, encroaching on protected forests. Lack of land ownership and appropriate alternative technologies, as well as general economic hardship make the practice of shifting cultivation an important conservation issue. It appears, however, that the rate of deforestation caused by uncontrolled logging is a much greater problem in Kachin State than that caused by shifting cultivation.

Uncontrolled mining for gold, jade and iron is another major conservation issue pointed to by stakeholders.[1] The jade from Phakant, in Kachin State, is known for its high quality. Before the cease-fire agreement between the SPDC and armed minority groups in 1994, most of the jade mines were controlled by minority group armed forces. After the cease-fire agreement, the SPDC had more control over jade-mining than the KIO and the NDAK. These groups receive financial and technical backing from tycoons in Hong Kong, Taiwan and China, and monopolise jade-mining, excluding small-scale business activities run by local people.

Chinese businesspeople are also turning an eye towards iron deposits in Kachin State. A previous research trip revealed evidence of piles of unprocessed iron ore stored in Customs warehouses in Yunnan Province. This mineral resource is a new item becoming popular in China's market, as it is a useful raw material for heavy industry development. It is anticipated that the impacts of extracting iron from Kachin State will be no less significant than those caused by gold and jade-mining.

A local NGO worker revealed that he hardly saw any significant economic gains from the above-mentioned mining activities for local people. The rapidly changing landscape of jade mines and the obvious impacts to the environment (such as the blockade of waterways, permanent human settlement and exploitation of forest resources) have been recorded. Although no environmental information or scientific reports about the impact of small-scale gold-mining are available locally, all local stakeholders realise it is a critical issue that they do not have the means to solve alone.

After the cease-fire agreements with armed groups in Kachin, the government extended its army bases throughout the state. Consequently, land confiscation and land clearing became common practice. Exploitation of forest resources to financially support the extension of these military units has caused great confusion and conflict over resource management. Apart from the government army units, other armed groups are also present in Kachin State, including the KIO, NDAK and splinter groups. These groups have substantial business interests in their demarcated territories; their presence and active involvement in resource exploitation pose significant additional threats to conservation and development.

Senior military officers are involved in timber and mining businesses. It is difficult to understand the current political boundary between military officers and armed minority groups, most of whom focus on businesses that make large short-term profits. It is unknown how much money they are making from the exploitation of natural resources and what proportion, if any, is being channelled into development projects for Kachin people.

Traditionally, hunting was a valued occupation for the Kachin, with animal trophies garnering respect for male hunters among their local communities. This traditional practice has become a conservation problem, as killing wildlife has been made easier due to readily available arms supplied by armed groups, and because there are highly attractive market prices for such products at the Myanmar–China border.

The largely illegal trade occurs mainly with China and Thailand and is a major cause of the depletion of wildlife populations within and outside existing Protected Areas (Rabinowitz et al. 1995; Martin 1997; Martin and Redford 2000). Rao et al. (2002) reported that hunting was the most serious threat to the long-term survival of wildlife in Myanmar's Protected Areas. They concluded that hunting beyond subsistence levels occurred throughout Kachin State, and seriously affected the whole wildlife population.

Local stakeholders reported that wood was still available but was becoming scarce in Kachin State. Charcoal use is traditional and continues widely. Apart from charcoal and wood, no alternative energy sources are conveniently available, especially in rural areas. Even in the large cities of Myitkyina and Bamaw, wood and charcoal are used predominantly for cooking as the electricity supply is unreliable. As long as the country's energy supply is inadequate, dependence on naturally available resources such as wood will remain high. Kachin State is no exception.

Non-timber forest products, including orchids and medicinal plants, are being collected in Kachin State to supply markets in neighbouring countries. Although there are strict regulations on their collection, enforcement is too weak to stop the illegal collection, transportation and marketing of these products. Most of the border markets in Yunnan Province are trading grounds for tremendous amounts of wild animal and plant products, collected mainly in Kachin State.

Infrastructure development in Kachin State is another potentially serious issue if these activities are not well regulated and monitored. It is apparent that road construction works implemented by Chinese contractors are not accompanied by environmental impact assessments (EIAs) and are not required to follow any environmental regulations. For example, part of the famous Ledo Road is now being upgraded in Kachin State to reconnect the road system between India and China. The previous road alignment was on high terrain with steep slopes, but the contractors chose a cheaper alignment along the waterways. Instead of constructing a proper drainage system for the removal of earth, they dumped it into creeks. Such negligence is common in all construction works.[2]

Fishing along waterways is a typical livelihood practice in Kachin State but has consequences for conservation because of the increasing use of environmentally unfriendly methods of catching fish. People involved in logging, mining and road construction works rely on fish as a major food source and often use dynamite and chemicals to catch a maximum amount of fish with minimal effort and in a short time (Pan Kachin 2004). These practices have serious environmental and social consequences for people living downstream. The traditional subsistence-fishing livelihoods of local people have been placed in jeopardy (Images Asia and Pan Kachin Development Society 2004).

Possible solutions and policy implementation

The current study involved discussion with various major donors in Myanmar about conservation issues in Kachin State. A priority for the donors was humanitarian assistance, but they were agreed that natural resource exploitation was a serious issue. Not taking timely action will result in irreplaceable losses for future generations. There is a window of opportunity at present if donors integrate an environmental component into their mainstream humanitarian programs. A meeting with senior officials from relevant ministries was organised, and they agreed that the current issues were significant and they expressed a willingness to tackle them. Commitment from the government is critical to Myanmar's conservation issues.

Ultimately, major political reform is essential to address conservation issues in Myanmar. Successive military councils have ruled the country for more than 40 years. The current regime has been in power as the State Law and Order Restoration Council (SLORC) since 1988 and the SPDC since 1997, and has granted many concessions to foreign investors for the purposes of extracting natural resources. Due to a lack of transparency and accountability, illegal practices have occurred across every level of resource exploitation. Unless transparency is improved enormously, reckless resource exploitation will continue.

Opposition groups inside and outside the country must unite in their preparation of effective strategies and alternative plans. The successive military governments of Myanmar have shown no inclination to bow to sanctions or other forms of international pressure. All indications suggest that the SPDC will maintain its grip on power in Myanmar by forging closer ties with China and India through natural resource deals such as those for natural gas and timber. The SPDC earns large amounts of revenue from these deals, which prop up the regime and provide all-important hard currency.

A Burmese academic who wants to remain anonymous has proposed that the regime's reluctance to engage in genuine political reform is due to fear of its own people. After intentionally creating entire systems—particularly in education, health and the economy—to favour the armed forces, the SPDC is concerned that there will be a revolution driven by its people and/or even by the army. Only by genuinely reforming the entire political system will attempts to address conservation issues be effective. The presence of many armed ethnic groups, causing complex governance in Kachin State, further demonstrates the urgent need for true political reform.

The State is the sole owner of Myanmar's natural resources, and so state institutions at various levels have the power to manage them. During the colonial days and for a short period after independence, natural resource institutions were well equipped with professionally trained staff and proper policies, regulations, manuals and instructions. At that time, professional staff had a certain degree of independence to implement their duties and take due responsibility.

The institutions responsible for natural resource management have, however, been more or less militarised in Myanmar. The militarisation of civilian and professional institutions has caused a major 'brain drain' of trained staff. The natural resource management institutions based in Kachin State are no exception. The situation is even worse in remote areas, where it is difficult to monitor institutional activities. An upgrading of the capacity of institutions through reform is urgently required.

As sole owner of the country's natural resources, the government even declares its ownership of areas not under its control. After the peace agreements with armed Kachin groups, the first permission granted to those groups was to exploit natural resources, particularly forests and mines. The rights of indigenous people to access natural resources and the small-scale business opportunities of local people have been largely ignored in this process. This lack of true ownership has compounded livelihood issues such as inequitable distribution of benefits within the country and the transfer of livelihood benefits outside the country. The obvious examples in Kachin State are the funds gained from natural resources that have been used largely for military spending by insurgent groups to fight the SPDC. Élites tend to be the main beneficiaries, while local communities continue to lack electricity, roads and other basic infrastructure. Roads built by logging companies are often fragmented and/or do not meet local needs, and logging companies are staffed by Chinese workers only, offering no employment opportunities for locals.

There is low customs compliance due to the regime's lack of control over areas serving China, and rampant corruption among staff. Revenue loss from illegal forest activities close to the Chinese border will continue to be high unless customs management is improved and coordinated with other agencies.

Addressing conservation issues usually transcends the political boundaries of individual countries and demands strong cooperation between governments. In November 2005, a joint committee between the governments of Myanmar and Yunnan Province was formed to combat illegal logging along the China–Myanmar border. Besides law enforcement, a range of opportunities should be opened up to include local people in wood-based industries, nature tourism and academic research.

Non-compliance of concessionaires and issues of concession management are among the driving forces of illegal forest activities occurring along the border area. Short logging contracts (some are less than one year) with Chinese companies promote poor management and reckless, shortsighted actions to tap whatever resources possible within a limited time. Reasonable long-term concessions with attached conditions to protect the environment, natural forest management and reforestation will be useful—unless a total logging ban is feasible.

Overseas development assistance to Myanmar has declined in the past 20 years. Less than 1 per cent of total overseas development assistance is used in general environmental protection and, compared with her five neighbouring countries (China, India, Laos, Thailand and Bangladesh), Myanmar receives the smallest amount of such assistance. Without substantial funding from external sources and rapid, genuine political reform, the natural resources of Kachin State will continue to be exploited in the name of development. The fact that environmental assistance is equally important to humanitarian assistance in Myanmar has been discussed in a number of online articles (Thaung 2003, 2004, 2005). It is time for donors to review their current policies and integrate environmental components into mainstream programs.

A first step to address illegal forest activities in Kachin State has been to organise a meeting among stakeholders. In November 2004, the Wildlife Conservation Society (WCS) held a workshop to identify the issues involved in establishing the world's largest tiger reserve in northern Myanmar. It drew a gathering of senior government officials, minority group organisations, the United Nations Development Programme (UNDP) and several international NGOs working in Kachin State. This kind of model would be useful in addressing conservation issues at a smaller level (Kachin State or northern Myanmar). The joint committee between forestry officials of Myanmar and China formed to monitor logging activities at the border, as mentioned above, should further develop strategic frameworks to tackle conservation issues in the state and should encourage the participation of stakeholders from all walks of life.

Because Kachin State is vast and ecologically diverse, the intended conservation models should cover landscape scale with the concept of integrated development. There is an urgent need for assistance from international conservation and development agencies. The WCS is the most prominent conservation NGO; it has already helped to establish five Protected Areas, three of them in Kachin State. With limited funding, the WCS alone is struggling to address the complex conservation issues in the state.

It is important to consider the traditional practices, values and rights of local people when addressing shifting cultivation, hunting, wood collection and production of non-timber forest products. The

Table 12.5   Official logging companies in northern Myanmar

Name	Teak	Other	State	Township
Dagon Timber	10,000	35,000	Kachin	Bhamaw
Shwe Mote That	2,000	7,000	"	Myitkyina
Century Dragon	3,000	20,000	"	Bhamaw
Glory Trading Co.	50,000		"	Bhamaw
Jade Land Co.		10,000	"	Bhamaw
Myat Noe Thu	3,000	15,000	"	Myitkyina
Lucre Wood Co.		8,000	Shan	Lashio
Lucre Wood Co.		15,000	Shan	Shwe Li
U Saw Paw		2,000	Kachin	Bhamaw
One Star Co.	3,500	20,000	Shan	Moemeik
Mo Min Tan		25,000	Shan	Moemeik
Ten Ways Co.		8,000	Shan	Lashio
Htoo	15,000		Kachin	Bhamaw
Htoo	5,000	15,000	Shan	Shwe Li
MTE	4,000		Kachin	Myitkyina
MTE	2,000		Kachin	Bhamaw
MTE		2,000	Shan	Lashio
MTE	10,000		Shan	Moemeik
MTE	3,000		Shan	Shwe Li
Total allowed timber in Hoppus Ton	110,500	182,000		

Source: Myanmar Forestry Department

relationship of local people to their natural environment and the ways they can participate actively in managing it should be better understood. In our previous report, the critical need for more research in this area was stated. Only through the active participation of all stakeholders will this great conservation task be accomplished.

The private sector is an important player in natural resource extraction, but can be a useful source of partners for conservation and development. The private sector in Myanmar and China is taking an important role through the gravitation of small-scale producers toward niche markets, where they can find comparative advantage by taking advantage of new and growing markets, new partnerships to supply capital, new technologies to lower the cost of sustainable production, and better organisation and empowerment of local producers. Table 12.5 lists national private timber concessionaires that are extracting timber in Kachin State. Together with their Chinese counterparts, their investment role in conservation and the community is great.

Conclusion

Kachin State is rich in natural resources. Its location near resource-hungry China and its rule by people in need of hard currency has resulted in the unsustainable exploitation of its natural resources. In addition, the complex governance system makes management of these resources difficult. This research has attempted to reflect the situation of the many voiceless people in Kachin State. A pragmatic approach is required to work together with all stakeholders. An opportunity should be opened for the active participation of local stakeholders in managing their resources not only for current but future generations. Regardless of the country's political situation, international assistance for conservation in Myanmar is needed urgently. Such aid is required not for the support of undemocratic practices, but to help the people of Myanmar, who deserve to manage their environment through the country's democratisation process.

Notes

1 Pan Kachin, a Kachin Development Association, prepared a comprehensive report about gold-mining activities in Kachin State. It included key players, types of mining and impact on the environment and livelihoods (Images Asia and PKDS 2004).
2 Personal observation through visits to Kachin State in March 2005 and January and July 2006.

References

Global Witness, 2005. *A Choice For China: ending the destruction of Burma's northern frontier forests*, Global Witness, London.

Images Asia and Pan Kachin Development Society, 2004. *At What Price: gold mining in Kachin State, Burma*. Nopburee Press, Chiang Mai. Available from http://www.ibiblio.org/obl/docs/gold%20pdf1.pdf.

Kahrl, F., Weyerhaeuser, H. and Yufang, S., 2004. *Implications for forests and livelihoods: navigating the border: an analysis of the China–Myanmar timber trade*, report prepared for Forest Trends and World Agroforestry Center, Washington, DC. Available from http://www.forest-trends.org/documents/publications/Kahrl_Navigating%20the%20Border_final.pdf.

Katsigris, E., Bull, G., White, A., Barr, C., Barney, K., Bun, Y., Kahrl, F., King, T., Lankin, A., Lebedev, A., Shearman, P., Sheingauz, A., Su, Y. and Weyerhaeuser, H., 2005. 'The China forest products trade: overview of Asia-Pacific supplying countries, impacts, and implications', *Forest Trends*. Available from http://www.forest-trends.org (accessed 15 April 2006).

Khun Sam (2006) 'Opium crops on the rise in Kachin State'. Available from http://www.irrawaddy.org (accessed 1 September 2006).

Martin, E. B., 1997. 'Wildlife products for sale in Myanmar', *Traffic Bulletin*, 17:33–44

——, and Redford, T., 2000. 'Wildlife for sale', *Biologist*, 47:27–30.

Olson, D. and Dinerstein, E., 1998. 'The global 200: a representation approach to conserving the earth's most biologically valuable ecoregions', *Conservation Biology*, 13(2):502–15.

Rabinowitz, A. Schaller, G. B. and Uga, U., 1995. 'A survey to assess the status of Sumatran rhinoceros and other large mammal species in Tamanthi Wildlife Sanctuary, Myanmar' , *Oryx* 29(2):123–28.

Rao, M., Rabinowitz, A. and Khaing, S.T., 2002. 'Status review of the Protected-Area system in Myanmar, with recommendations for conservation planning', *Conservation Biology*, 16(2):360–7.

Thaung, T., 2003. 'A personal view: community forestry in Myanmar', *Community Forestry E-News*, No.16 (October), Regional Community Forestry Training Center for Asia and the Pacific, Bangkok. Available from http://www.recoftc.org (accessed 31 October 2003).

——, 2004. 'Protecting Burma's biodiversity in the long term', *Irrawaddy online commentary*, 30 March. Available from http://www.irrawaddy.org (accessed 30 March 2004).

——, 2005. 'Ten years on community forestry in Myanmar', *Community Forestry E-News*, No.10 (October), Regional Community Forestry Training Center for Asia and the Pacific. Available from http://www.recoftc.org (accessed 15 November 2005).

Webb, E., Thiha and Thaung, T., 2004. *Conservation issues of Kachin State, Myanmar*, phase 1, Report to MacArthur Foundation, May–July.

Acknowledgments

This trip was funded by the John D. and Catherine T. MacArthur Foundation and was conducted in collaboration with previous working partners in Kachin State, Myanmar. Thanks to David Hulse from the MacArthur Foundation and to Associate Professor Dr Edward Webb and PhD candidate Thiha from the Asian Institute of Technology for their support and constructive comments on the manuscript. Previous colleagues from SWISSAID Myanmar Program, Ja Tum Seng and G. Zung Ting, provided invaluable logistical support for the trip and interviews. In addition, thanks are due to local businessman U Chit Wai, from Phakant jade-mining town, and U Sein Win, from Muse, for their input to the interviews and networking for future trips.

Index

agriculture 9, 43, 44, 65, 68, 69,
  110, 112, 129–30, 132n, 175, 176,
  192–3, 253
  freedom to select crops grown 9, 10,
  44, 130, 201, 264
  labour force 129, 193
  Nyaunglebin District 251
  see also physic nuts
anti-corruption 2, 9
Anti-Fascist People's Freedom
  League 30
armed groups 7, 11, 12, 41, 60, 66–7,
  72, 224, 278, 280, 283
  Kachin State 275
army
  see military
ASEAN Inter-Parliamentary Union
  Myanmar Caucus 85, 103n
Asia Wealth Bank 128
Asia-Europe Meeting (ASEM) 91–2,
  104n
Asia-Europe Summit 84
Asian 'tiger' economies 112
Asian Development Bank 15–16, 111,
  213
Asian highways 173, 174
Association of Southeast Asian Nations

(ASEAN) 5, 82, 84–7, 92, 101, 102,
  103n, 104n
  and Aung San Suu Kyi 85, 86, 101
  environmental issues 205, 211, 213
  Regional Centre for Biodiversity
  Conservation 205
  Secretary-General 86
  United States involvement 85, 91
Atlantic Charter 30
Aung San Suu Kyi 11, 23, 212
  ASEAN support for 86, 101
  attacks against 1, 7, 8, 183n
  Europe support for 92
  house arrest 23, 86
  Japan support for 90
  lifting of restrictions on 23
  political challenge by xvii
  relationship with military 23, 24,
  29, 37
  UN support for 94, 99
  see also National League for
  Democracy (NLD)
authoritarianism 26, 27, 29–30, 31,
  34, 191, 218, 223

Bangladesh 123, 163
banking 113–15, 126–8

crisis 2002–03 xvii, 109, 127–8
  lending 114, 115, 126, 128, 129
bilateral cooperation 86–7, 165,
  175–6, 182n
bio-diesel/fuel 10, 37, 49n
Bo Mya 60
border areas 44, 45, 49–50, 50n, 56–
  7, 60, 66, 162–3, 165, 172, 174–5,
  178, 182n, 213, 280, 281, 284
British era 30, 45, 59, 146
Buddhism 59, 60
Burma Selection System (BSS) 194,
  215n
Burma Socialist Program Party
  (BSPP) 19, 20, 32
Bush, George 90, 91

Cambodia 125, 126, 162, 192
  workers abroad 171, 172, 177, 180
Canada 219, 227
capital city, official relocation xvii, 2,
  5–6, 95, 119
cease-fires/cease-fire groups xviii, 3, 4,
  7, 12, 13, 14, 20, 21, 37, 41–2, 57,
  61, 62, 67, 68, 69, 70, 71
  SPDC-KNU 63, 64, 65, 66, 254,
  265n, 279, 283
census 13, 194
Cental Bank of Myanmar (CBM) 113,
  114, 115–17, 127
Central Bank of Myanmar Law 126
children 75, 130, 170, 172
Chile 29–30
Chin State 56, 124, 150
China xviii, 11, 43, 50, 70, 89, 103n,
  105n, 112, 162, 163, 181n, 283,
  284, 286
  demand for resources 271, 276–7,
  279, 280, 287
  investment in Myanmar 50n, 87,

112, 123, 165, 182n
  political support/influence 87, 88,
  101, 102
  relationship with Khin Nyunt 45
  trade 102, 104n, 115
Chinese employers 177, 178, 184n,
  224, 240n, 276
civil conflict 20, 31, 32, 41, 45, 54–78
civil service xvii, 8, 9
  corruption 2, 9, 47
  rice supply 136, 142, 153, 156n
  salaries 9, 47, 115, 116, 119, 176
civil society 7, 67, 69, 71, 75
Commonwealth, the 30
community-based organisations
  (CBOs) 65, 71, 73, 75
Conservation International 212
constitution draft 1, 3–5, 6–7, 8, 14,
  21
  chapters 4
  future state structure 21, 32
  military phase-out, lack of 15
Constitutional Tribunal 5
corruption 2, 9, 10, 25, 26, 39, 60,
  103n, 110, 115, 119, 122, 195, 253,
  275, 277, 278, 284
  see also military regime, rent–seeking
coup, military
  March 2002 23, 24
  1988 49, 189

Daewoo International
  Corporation 123, 124, 192, 232
decentralisation 21–2, 32, 34
Defence Services Intelligence 26
democracy 18–34, 143, 182n
  activists 26
  consociational 21, 34n
  earlier period 19, 206
  missed opportunities 28–9

1988 uprising 10, 16, 19, 28, 29, 33,
  60, 136, 141, 142, 143, 223
  transition to 19, 20–1, 24, 210
Democratic Kayin Buddhist Army
  (DKBA) 60, 61, 265
Department of Mines 261, 262–3
Depayin Massacre xvii, 8, 76, 89,
  104n
developing-country status 19, 34
development assistance 22, 90, 104n,
  140, 141, 284
  see also foreign investment;
  humanitarian aid
displacement, people
  see migration, forced
drug trade 61, 70, 88, 91

Earth Summit Plus Five 189–90
EarthRights International 234, 235,
  236, 238
Economic Cooperation Strategy
  (ECS) 162–3, 181
economy xviii, 2, 34, 69, 108–31
  agriculture 110, 112, 129–30, 192–3
  centrally planned 20, 32
  exchange rate 115, 117–18, 119,
  131n
  exports 108, 112, 119, 120, 131n,
  136, 143–5, 146, 151, 156n, 181–2n
  FDI flows 122–3, 125, 223–4
  fiscal policy 46, 113–15
  foreign exchange 125–6, 135, 151,
  153, 222
  GDP 111, 112, 129, 192
  GDP growth rate (claimed) 108,
  110, 128
  growth 108, 110–13, 163
  imports 118, 120, 121, 131n, 181n
  import substitution 175
  inflation 109, 115, 116, 119, 129,
  234, 263
  interest rates 115–16, 127
  macroeconomic policy 109, 113,
  118–19
  manufacturing 111–12, 121
  monetary policy 115–17
  opening 47, 123, 189, 223
  private sector 111–12, 113, 114,
  126, 128, 145, 150, 151, 153, 161,
  286–7
  reform xviii, 88, 108
  services 112, 121, 170
  trade surplus 112, 119–21
education xviii, 73, 75, 110
elections 7
  1990 7, 12, 14, 19, 22, 29
  prospects for 12–14
  SPDC road-map 2, 5, 6
  voter lists 13
electricity 9, 111, 141, 149, 253, 281,
  284
  see also hydroelectricity; infrastructure
embassies 5–6
employment 166, 167, 255
  see also labour force
environmental conflict
  resolution 238–9
environmental issues xix, 70, 87–8,
  93, 189–215, 271–93
  agriculture sustainability 192–3
  chemicals and pesticides 193, 228,
  229, 230, 234, 252–3, 282
  conservation issues, Kachin
  State 271–83
  conservation areas 212
  deforestation 193–4, 209, 230,
  258–9, 271, 277, 278, 280
  ecological diversity 191–2
  environmental and social impact
  assessments 192, 196, 200, 225–6,

239, 281
fishing 179, 273, 275, 276, 281–2
governance 200–1, 210–13, 218–39
international conventions/
treaties 202–3, 204, 205, 206, 211
legislation 198–9, 200
local people's rights 204
national parks 208, 210
public awareness-raising/
education 195, 196, 210, 237
Smithsonian Institution ecological
study 207–9
soil degradation 193, 200, 228, 230,
234, 264
training government offices 211
water pollution 193, 228, 229–30
water and water resources
management 44, 73, 175, 193, 195,
200, 229, 230, 234, 252–3, 262,
279, 281
United Nations Development
Programme 209–10, 214
wildlife conservation 207–8, 208,
280–1
see also mining
ethnic groups 7, 11, 12, 21, 22, 191
armed groups 7, 12, 60, 278, 283
cease-fire groups 3
insurgent groups 59–60, 62
non-Burman 11
ethnic
Chinese 176
diversity 7
minorities/communities 12, 62, 94,
183n
Europe 91–3
and Aung San Suu Kyi 92
Southern 27
European Union 15, 91
sanctions 93, 104n

exchange rate 115, 117–18, 119, 131n
Extractive Industries Transparency
Initiative (EITI) 237

farming
see agriculture
federalism 7, 12, 32
Financial Action Task Force 129
Financial Institutions of Myanmar
Law 126
financial system 45, 109, 113–15,
126–8
deregulation 45
selected financial indicators 127
see also banking; economy
fishing 177, 179, 273, 275, 276,
281–2
law 198
forced migration
see migration, forced
foreign investment xviii, 16, 87, 110,
175, 189–90, 192, 200
Chinese 50n, 87, 112, 123, 165,
182n
direct 104n, 122–3, 125, 223–5
mining 110, 223–5, 227, 282
Myanmar Foreign Investment
Law 224
Thai 49–50n, 123–5, 130
foreign policy 82–102
Forest Conservation Committee 193
forestry 193–5, 209, 285–6
1992 Forest Law 194–5, 198
see also environmental issues; logging

garments and textiles 120, 121, 162,
164, 172, 179, 181n
joint ventures 159
Gas Authority of India Limited
(GAIL) 123

gas
  see natural gas
Global Environmental Facility 211
Global Fund for HIV/AIDS 77, 95,
  105n
globalisation 42, 45, 176, 204
Goh Chok Tong 86, 92
governance 19, 213–14
  environmental 190–1, 192, 195,
  200–1, 204–5, 206, 209, 212, 213,
  218–39
  lack of 190
  Nyaunglebin District 246–66
  'tandem' 29
Greater Mekong Subregion
  (GMS) 174, 213

health xviii, 73, 77, 95, 105n, 110,
  180–1, 185n, 251, 253
Heritage Foundation 122
heroin 69
HIV/AIDS 75, 181
  Global Fund for 77, 95, 105n
Hogyit Dam 165, 182n
Hu Jintao 88
human rights
  abuses xviii, 12, 15, 56, 63, 64, 67,
  68, 94, 103n, 204, 239
  advocacy 74–7
  draft guidelines for agencies 76–7
  humanitarian 'space' 76–7
  protection 72–3
humanitarian aid 68, 70, 83, 90, 94,
  95, 285
Humanitarian Dialogue 95
hydroelectricity 165, 182n, 192, 248,
  249, 252, 253, 254, 256, 258, 264

Ibrahim Gambari 99, 100, 105n
income xviii, 70

civil service salaries 9, 47, 115, 116,
  119, 176
  migrant workers 176, 177, 184n
  rises in per capita 19
India 11, 50, 82, 89, 112, 123, 124,
  163, 168, 283
Indonesia 15, 40, 86, 103n
industrialisation 126, 156, 159–81
  industrial zones 159, 160, 164,
  165–7
  private sector 161–2
infrastructure 9–10, 44, 69, 87, 89,
  121, 258, 278, 284
institutions 28, 30, 32, 33, 283
  development 195–6, 197, 210, 211,
  222
insurgency 42, 44, 54, 55, 59–60, 62,
  63, 67, 68, 71, 284
  counter-insurgency 63, 250, 251,
  254
  'four cuts' counter-insurgency
  strategy 63
internal displacement
  see migration, forced
international agencies 62, 72, 73, 74,
  75–7, 95, 95, 96
international assistance/aid xviii, xix,
  13, 211–12, 285
  for conservation 284–5, 287
  see also humanitarian aid
International Committee of the Red
  Cross (ICRC) xviii, 72, 74–5, 96,
  105n
international community xviii, 3, 5,
  13, 15, 83, 91, 100–1, 210, 211
International Covenant on Economic,
  Social and Cultural Rights 72
International Labour Conference 96,
  97
International Labour Organization

(ILO) 83, 96–8, 105n, 205
International Monetary Fund
  (IMF) 16, 118, 131n
international relations 82–103
  policies 93–100
Iraq 13, 84
Irrawaddy-Chao Phraya-Mekong
  Economic Cooperation Strategy 175
Ivanhoe Mines Limited 219, 227, 228,
  230, 231, 232, 233, 237, 240n

jade 68, 279
Jakarta Declaration on Environment
  and Development 205
Japan 89–90, 91, 104n, 140, 227–8
  and Aung San Suu Kyi 90

Kachin Independence Organisation
  (KIO) 67, 68, 279, 280
Kachin State 7, 67, 68, 71
  armed groups 275
  case study 271–87
  conservation concerns 273–4,
  275–82
  conservation solutions, possible 282–
  7
  energy sources 281
  fishing 280–1
  hunting 280–1
  infrastructure 278, 284
  international assistance 285
  mining, uncontrolled 279
  opium 277
Karen National Liberation Army
  (KNLA) 61, 251, 252, 258, 265n
Karen National Union (KNU) xviii,
  12, 59–66, 71, 174, 251
  conflict 248, 251, 252, 264
  KNU-SPDC cease-fire 63, 64, 65,
  66, 68, 254, 265n, 279, 283

Karen State 77, 163
  case study 59–63, 65
Khin Nyunt xvii, 11, 37, 63, 95, 171
  ambition 25, 26
  and Aung San Suu Kyi 23
  foreign policy 83, 84, 85
  power base 25–6
  relationship with China 45
  relationship with Ne Win 24, 26
  2004 sacking 37, 76, 82
Kofi Annan 98
Kokang 45, 69–70
Korean Gas Corporation 123
Korean Resources Corporation 232
Kyaut Nagar Dam 253–4, 256,
  257–8, 264, 266n
kyay-zu ('good deeds') 1–2, 40, 41,
  42–4, 46
kyet-su
  see physic nuts

labour/labour force 167–8, 248
  agriculture 129, 193
  contribution fees 44
  exported 49n, 57, 168, 254
  forced 9, 43, 44, 56, 64, 67, 69,
  96–7, 105n, 252, 256, 258
  illegal migrants in Thailand 169,
  170–2
  numbers in Thailand 182n
  'slave workers' 176–7
  women 170
  see also employment
land
  confiscation 67, 68–9, 257
  leasing to private companies 68
  under cultivation 130
Laos 162, 192
  workers abroad 171, 172, 178, 179
Latin America 27, 181n

legislation 91, 278
  Central Bank of Myanmar Law 126
  environmental 194–5, 196, 197,
    198–9, 200, 211, 240n
  Financial Institutions of Myanmar
    Law 126
  Foreign Investment Law 224–5
  1992 Forest Law 194–5, 198
  1994 Mines Law 225, 226
  mining 223–6
living standards xviii, 46, 204
logging xix, 61, 194, 271
  companies in Northern
    Myanmar 286
  illegal 87–8, 121, 192, 194, 196,
    271, 277–8
  Kachin State 277–8, 279, 284
  see also environmental issues; forestry
London Metals Exchange 227, 233

Malaysia 86, 168, 185n
martial law 191, 214n
Marubeni Corporation 227–8
Maung Aye 23, 37–8, 83, 93
Maung Maung Swe 38
Médécin sans Frontières (MSF) 77, 95
media 9, 10–11, 85, 95, 104n, 170,
  184n, 191
Mekong River Commission
  (MRC) 205, 206, 211
migration, forced 54–78, 251, 254,
  264
  armed conflict-induced (Type
    One) 55–7, 72, 251
  internal displacement, population
    estimates 57–9
  Karen, the 59–63, 65, 250, 251, 254
  livelihood vulnerability-induced
    (Type Three) 56–7, 67, 69–70
  Military Affairs Security 38, 49n

rehabilitation 65–6, 67, 72
  state-society conflict-induced (Type
    Two) 56, 57, 58, 67–9
  three types of displacement 55–7
  typology 56
military (Tatmadaw) 3, 4, 7, 15,
  36–48, 56, 58, 59–73, 130, 163,
  174, 235, 249–50, 251, 252, 253,
  254, 255–6, 258, 259, 260–2, 264,
  280, 283
  armed forces, links with the past 24–
    7
  see also military regime; State Peace
    and Development Council
Military Affairs Security 38, 49n
military coup
  see coup, military
Military Intelligence xvii, 11, 26, 49n,
  63, 171
military regime xvii, xviii, 1, 8, 10, 11,
  12, 20, 23, 34, 36–50, 77, 83–4, 97,
  102, 109, 176, 180
  cars 40, 41, 49n
  corporate ventures 43, 44–5,
    109–10, 232
  current 30, 31
  demobilisation 48
  expansion 43–4, 46–7, 163
  internal conflict 37, 39, 48
  lack of general support for 8–9
  participation in Parliament 3, 15, 27
  regional commanders 38–9, 43
  relatives of officers 45–6
  rent-seeking 39, 42, 44, 109–10,
    226, 252, 256, 259–62
  self-reliance 46–8, 130
  wealth/poverty 40–1, 42–4, 47
  weaponry/military equipment 47,
    49n
  see also military (Tatmadaw); State

Peace and Development Council
Min A Naw Ya Ta 253, 258
Min Ko Naing 11
mining 60, 61, 218–39
  access to information 221
  artisanal 219, 220, 234–5, 237–8
  ASM 219, 220, 237
  Chinese companies 224, 240n, 276
  deaths 234
  equipment needs 233
  finance 224–5, 228, 231, 232, 233
  gold, 248, 249, 250, 252, 254, 256,
  257, 258, 259, 260–1, 262, 279, 280
  governance, local participation 237,
  238–9
  hydraulic mining machines 262,
  266n
  iron ore 221, 279
  jade 68, 276, 279
  Korean consortium 232, 233
  legislation 223–6, 227
  1994 Mines Law 225, 226, 227,
  262–3
  overview 219–21
  Peace and Development Council 253
  precautionary principle 235–6, 240n
  price 262
  responsible 235–9
  voluntary agreements 230–1, 237
  see also Monywa Copper Project;
  Shwegin Township
Ministry for Progress of Border
  Areas 44
Ministry of Electric Power
  Enterprise 253
Ministry of Finance and Revenue 116,
  119
Ministry of Home Affairs 94–5
Ministry of Mines 220, 222, 227,
  240n

Ministry of National Planning and
  Economic Development 76, 116,
  223
Mon State 7, 16n, 67, 68–9, 70, 77,
  162
money laundering 91, 128–9
Monywa Copper Project 218–19, 219,
  224, 227–33
  artisanal 219, 220, 234–5, 237–8
  deaths 234
  environmental management
  system 230–1
  expansion 231–2
  joint venture agreement 227
  methods and processes 228–30
  security 235
Multi-Fibre Agreement 121
Myanmar (editors' note) xvi
Myanmar Agricultural Development
  Bank 129
Myanmar Agricultural Produce Trading
  (MAPT) 140, 152, 153
Myanmar Economic Holdings 25
Myanmar Electric Power
  Enterprise 253
Myanmar Foreign Investment
  Law 224
Myanmar Industrial Development
  Committee 161
Myanmar Ivanhoe Copper Company
  Limited (MICCL) 227, 230, 232,
  235
Myanmar Mayflower Bank 128
Myanmar Ministry of Mines 219, 224
Myanmar National Committee for
  Women's Affairs 167
Myanmar National Democratic
  Alliance Army (MNDAA) 69, 70
Myanmar Oil and Gas Enterprise
  (MOGE) 123, 124

Myanmar Women's Affairs Federation
   (MWAF) 8, 13, 97

Naypyitaw xvii, 3, 5, 7, 94, 118, 162
Naypyitaw Command 27
National Commission on
   Environmental Affairs (NCEA) 193,
   195–6, 200, 211–12, 215n
National Convention 1–2, 3–5, 7, 14,
   103n
   'Group of Eight' 3–4
   military involvement 15
   resumption of xvii, 1, 2, 3, 99
National Environment Policy 221–3
National Environmental Action
   Plan 195, 196–200
National League for Democracy
   (NLD) xviii, 1, 10, 11, 14, 19, 99,
   101, 190
   approach to democracy 20
   attacks by SPDC 7, 11, 12, 15, 29
   continuation as a party 22
   leadership in detention xviii, 3, 85,
   90
   1990 election 14
   see also Aung San Suu Kyi
National Registration Cards 45
natural gas 82, 108, 112, 119, 120,
   121, 182n
   foreign investment 110, 122, 123
   new ventures 123–5
   pipeline 192
natural resources xix, 39, 45, 48, 61,
   67, 131, 248, 275
   ownership 283
   see also mining; natural gas
Ne Win 19, 23, 24, 59
   links to others 24, 25
   relatives 24
New Mon State Party (NMSP) 67, 68

non-government organisations
   (NGOs) 65, 71, 73, 129, 206, 231,
   236, 279
   international (INGOs) 68, 76, 83,
   94–5, 105n, 191, 206, 211, 230, 285
North Korea 93, 94, 104n, 105n
nuclear research assistance 93, 104n
Number One Mining Enterprise
   (MEI) 227, 240n
Nyan Win 83
Nyaunglebin District 246–63
   description 250

oil 192
   foreign investment 110, 122, 123,
   124
Olympic Company Limited 253, 258
Ong Keng Yong 86
Oil and Natural Gas Corporation
   (ONGC) Videsh 124
opium 57, 67, 69–70, 73, 277
Organisation for Economic
   Cooperation and Development
   (OECD) 129
overseas development assistance
   see international assistance

Panglong Agreement 21
peace agreements
   see cease-fires
people trafficking 75, 91, 170, 172,
   176
PetroChina 124
Petronas 125
physic nuts (kyet–su) 10, 37, 42, 49n
'Pinheiro Principles' 71, 78n
police 24, 26, 44, 75, 172, 176
   Thai Immigration 178, 180, 184n
political
   activists, expatriate 85

prisoners 96, 105n
reform 39, 85, 103n, 282, 283, 285
political parties 29, 31
  democratic opposition xvii, 20, 21,
  22–4, 34, 212
  main opposition 20, 22
  'organised' opposition 11–12, 14, 26
  see also name of party
politics
  Bamar (majority) 21, 32
  lack of interest in 14
  military involvement in 3, 15, 27
poverty 19, 20, 32, 40, 41, 69, 109,
  111, 234, 257, 263
Premier Oil 125
prostitution
  see sex trade
PTT Exploration and Production 124,
  125
Pyinmana
  see Naypyitaw

Rakhine State 121, 123, 124–5, 150
reconciliation, national 7, 82, 85, 86
refugees 57, 61, 65, 68, 71–2, 170,
  183n
  see also migration, forced
renewable energy 10
rice 65, 70, 73, 130, 146, 147, 155
rice marketing 135–56
  export 136, 143–5, 151, 153
  export deregulation 136, 143, 153–4
  export destinations 143, 144
  first liberalisation 136–45
  government intervention 149–50,
  153
  milling 140–1, 145–6
  milling fee 147–8
  mill problems 146–9
  paddy procurement system 135, 136,

136–40, 143, 148, 152
  prices 137, 140, 143, 145, 150, 152,
  156, 156n
  private sector 135, 140–1, 143, 145,
  147–8, 151, 153
  public service supply 136, 142, 153,
  156n
  quality 140, 143, 153, 155–6
  quotas 151
  rationing 135–6, 141–3, 145, 151,
  152–3
  second liberalisation 151–4
  volume 138–9, 145
  wholesalers 146, 151
Rio Declaration 236
rural areas
  armed conflict 59
  displacement 63
  fuel sources 194, 281
  population 9, 14, 16n
  poverty 130, 132n, 204
  SPDC support 7, 9, 14, 16n
Russia 93, 98

Salween River 64, 192
sanctions 100, 282
  bilateral 84
  Chinese opposition to 87
  European Union 93
  international 98, 103n
  United States 117, 162, 232–3
  West, the 20
Saw Bo Mya 60
Saw Maung 23
sex trade 170, 176, 181, 183n
Shan Nationalities League for
  Democracy xviii
Shan State 73, 150, 163, 164, 209
  Army (South) 11, 12, 174
Shwe Gas 232, 240n

Shwegyin Township 246, 249, 250–64
  gold 250, 252–3, 254, 256, 257,
  258, 259, 267, 264
  hydroelectricity 252, 253, 254
  mining area 247
  rents extracted 260–1
  rubber plantations 253, 264
  state-sponsored violence 255–6
  waterways 251–5
Singapore 86, 92, 104n, 123
  foreign workers 178
Smithsonian Institution 207–8, 209
smuggling 60, 163–4, 176
Soe Tha 110
Soe Win 82, 83, 88
South Korea 112, 123, 181n, 232
  foreign exchange reserves 125
State Law and Order Restoration
  Council (SLORC) 42, 48, 223, 282
State Peace and Development Council
  (SPDC) 1–16, 37, 48, 61, 73, 88,
  93, 108, 111, 112, 113, 117, 128,
  129, 282–3
  attacks on political challengers 10–
  11, 15, 84, 201
  and Aung San Suu Kyi 86
  handling of coup attempt 25
  military coup 189
  opposition to federalism 7, 12
  relationship with ASEAN 86
  relationship with NLD 7, 11, 12, 15,
  29, 37
  road-map xviii, 1, 2, 5, 6, 8, 12, 13,
  14, 82
    seven steps 2
  SPDC-KNU cease-fire 63, 64, 65,
  66, 68, 254, 265n, 279, 283
  see also Maung Aye; military regime;
  Than Shwe
state structure, future 21, 32

students 1, 10, 16n, 232
sugar cane 162, 175
Susilo Bambang Yudhoyono 86
Syed Hamid Albar 86

Taihan Electric Wire 232
Tatmadaw
  see military
taxation 43, 44, 64, 113, 165, 167,
  252, 257
  black-market goods 60, 61
  customs duty revenues 113, 114
  mining 233
  proportion of government
  spending 115
teak 120, 121, 286
textiles 120, 121, 162, 164, 172, 179,
  181n, 159
Thailand 15, 29, 43, 45, 60, 74, 86,
  89, 102, 104n, 112, 113, 162, 163,
  280
  bilateral cooperation 86–7, 165,
  175–6, 182n
  Burmese illegal workers 170–2, 174,
  175, 180, 184n
  Burmese migrant workers 167–70,
  170–2, 176–8, 179, 184n
  deportation of Burmese
  nationals 171, 179
  economy 179
  education sector 179
  foreign exchange reserves 125
  Immigration Police 178, 180, 184n
  industrial zones along border 165–7
  investment in Myanmar 49–50n,
  123, 124, 130
  memorandum of understanding 130,
  183n
  Myawaddy-Mae Sot area 172–5,
  181, 181n

refugee camps 61, 65, 68, 170, 171
skilled workers 179
trade 115
workers abroad 177
Thailand Burma Border Consortium
(TBBC) 58, 64
Thaksin Shinawatra 86, 102, 171,
183n, 185n
Than Shwe 7, 9, 11, 37, 99, 101,
183n
appointment 24
and Aung San Suu Kyi 23
visit to China 88
visit to India 82
Thein Sein 3, 82, 103n
Total Oil 125, 192
tourism xviii, 82, 112, 200, 275
trade, illegal 213, 280

U Ohn Gyaw 204
U Soe Tha 104n, 116–17
U Thaung 83
unemployment 161, 167–70, 276
rate and labour force 168
UNICEF 72, 73, 75, 130
Union for Solidarity and Development
Association (USDA) 2, 8, 13, 14, 32,
44, 76, 96
political role 8, 10, 97, 183n, 184n
Union of Myanmar Economic
Holdings Limited (UMEHL) 43, 47
United Kingdom 98, 125
United Nations High Commissioner
for Refugees (UNHCR) 68, 72, 75,
76, 171
United Nations xvi, 70, 72, 74, 75–6,
76, 84, 94, 98–100, 101, 191, 194,
197, 212
Conference on Environment and
Development 205

Development Programme
(UNDP) 129, 209–10, 211, 214,
285
Environment Program (UNEP) 211
Inter-Agency Standing
Committee 76
Secretary-General 94, 98, 100
Security Council 87, 88, 89, 91, 93,
98–9, 100, 105n
Special Envoy 23, 83, 88, 94
Special Rapporteur for Human
Rights 83, 94
Sub Commission on the Promotion
and Protection of Human Rights 71
United States of America 7, 15, 60,
90–1, 98, 125, 181n
ASEAN activities 85, 91
sanctions 117, 162, 232–3
unilateralism 84
United Wa State Party 70
Unocal 125, 192
uprising, 1988 10, 16, 19, 28, 29, 33,
60, 136, 141, 142, 143, 223

Vietnam xviii, 163
foreign exchange reserves 125
violence 28, 39, 48, 73, 232, 248,
254, 255–6, 263
state sponsored 250–1, 255–6, 257
voter lists 13

Wa sub-state 70
War Office 38, 47
War Veterans Association 97
West, the xviii, 6, 8, 10, 12
sanctions 20
Wildlife Conservation Society 208,
209, 285
women 8, 25, 61, 97, 162, 167, 170,
204, 263